MW00779302

A Guide to
Australian
White
Cockatoos
THEIR MANAGEMENT, CARE
& BREEDING

By Chris Hunt

**First Published 1999 by
ABK Publications
PO Box 6288,
South Tweed Heads,
NSW. 2486. Australia.**

ISBN 0 9577024 18

Front Cover:
Top left: Eastern Long-billed Corella
Centre left: Major Mitchell's Cockatoo
Centre right: Sulphur-crested Cockatoo
Bottom left: Galah
Bottom right: Short-billed Corella
Back Cover: Western Long-billed Corella

All photographs by Chris Hunt except where shown.

Design, Type and Art: PrintHouse Multimedia Graphics (Gold Coast)
Colour Separations: Nu Scan (Gold Coast)
Printing: Prestige Litho (Brisbane)

CONTENTS

ABOUT THE AUTHOR

Chris Hunt was born in Bendigo, Victoria in 1956 and like many Australian youngsters, by the time he was ten years old, had a collection of Zebra Finches, Budgerigars and King Quail in the backyard. A plumber by trade, Chris was a professional cyclist of world standing for many years and actually held the Australian and Australasian one hour cycling record in 1984, just before he retired from the sport.

After marrying and moving to Morwell in 1985, Chris re-established his fascination in birds and began breeding Neophema Grass Parrots and various mutations of Cockatiels. During early 1992, Chris and his wife Maree relocated to rural Hazelwood in Victoria, which afforded him the opportunity to keep and breed larger parrots and cockatoos. Chris has kept and bred all species of Australian white cockatoos and currently maintains a collection of various white cockatoos, Red-tailed Black Cockatoos, Gang Gangs, Eclectus and Bush Budgerigars.

Together with Maree, Chris handrears many species of white cockatoo for the pet market. His wealth of experience is well respected throughout aviculture in this country.

A longstanding member of the Avicultural Society of Australia, Chris has also held the position of Vice-President of the Latrobe Valley Avicultural Club, was foundation President of the Australian National Cockatiel Society - Gippsland Branch, which eventually became the Victorian Cockatiel Society and is current President of the Avicultural Society of Australia - Gippsland Branch. Chris is a keen communicator and believes in assisting in the dissemination of information to other aviculturists via lectures and discussions. He is a regular speaker at various avicultural meetings in Victoria and contributes many articles to club journals throughout Australia and *Australian Birdkeeper Magazine*.

ACKNOWLEDGMENTS

My very special appreciation and gratitude to Diana Andersen for her help and patience, Dr Stacey Gelis BVSc (Hons) MACVSc (Avian Health) for his contribution and hard work on the Diseases and Common Disorders section, all members of the Avicultural Society of Australia - Gippsland Branch, Nigel and Sheryll Steele-Boyce, who gave me the opportunity and encouragement to write this book and who have also helped me immensely, my aunties Nancy and Doreen and last, but by no means least, my wife Maree, whose support has always encouraged me.

I would also like to sincerely thank the following people for their assistance in the preparation of this book.

Geoff Andrew, Pauline and Denis Cook, Kay Fender, Graeme Hyde, Ron Johnstone (University of Western Australia), Edda Kernstock for the drawings, Rosemary Low, Philip McRoberts Dip.App.Sc., BA (McRoberts & Sons Seed Merchants), Chris Osborne, Joy Pym, Andrew Young (Australian Bird Company, Melbourne), Dr Terry Martin BVSc for Galah Mutations - Chart of Genetic Expectations, Dr Danny Brown BVSc. (Hons), BSc. (Hons), Dr Bob Doneley BVSc., MACVSc. (Avian Health), Tod Osborne and Phil Digney.

FOREWORD

Australia's white cockatoos have fascinated the world for over 200 years, and their popularity is evidenced by the frequency with which you encounter them in Australia's backyards as pets. However, it has often been said that 'familiarity breeds contempt'. It is unfortunate that their proper care, both as pets and in aviculture, has long been neglected. If aviculturists truly wish to argue that they keep birds because they love them, this situation must be remedied.

Stan Sindel and Bob Lynn first addressed this in their book **Australian Cockatoos**. Chris Hunt has now taken it further in a superb text suitable for both pet owner and aviculturist. This is a well-researched, well-written book, with facets of personal experience shining throughout its pages. Richly illustrated and full of practical hints, it is sure to become a standard handbook for the breeders and keepers of these magnificent birds, both in Australia and overseas.

With the help of this book, aviculturists will be able to lift the standard of care for these birds and hopefully gain a new awareness of their uniqueness.

Chris Hunt and **ABK Publications** are to be congratulated for adding this book to the **'A Guide to...'** series.

Dr Bob Doneley BVSc MACVSc (Avian Health)

K. WILSON

Galah *C.r. roseicapilla* hen

What is a cockatoo? Cockatoos are defined by ornithologists as parrots with a movable crest and can only be found in Australasia, ranging in distribution from Australia, New Guinea, the islands of Indonesia, the Solomon Islands and the Philippines. There are 16 recognised species of white or pale-coloured cockatoos, seven of which are endemic to Australia. They range in size from 31cm to 50cm in length. All belong to the *Cacatua* genus, with the exception of the Cockatiel *Nymphicus hollandicus* and include the following.

Australian Species
Sulphur-crested Cockatoo *Cacatua galerita*
Short-billed Corella *Cacatua sanguinea*
Eastern Long-billed Corella *Cacatua tenuirostris*
Western Long-billed Corella *Cacatua pastinator*
Major Mitchell's Cockatoo *Cacatua leadbeateri*
Galah *Cacatua roseicapilla*
Cockatiel *Nymphicus hollandicus*

Other Species
Yellow-crested or Lesser Sulphur-crested Cockatoo *Cacatua sulphurea*
Citron-crested Cockatoo *Cacatua citrinocristata*
Triton Cockatoo *Cacatua galerita triton*
Blue-eyed Cockatoo *Cacatua ophthalmica*
Salmon-crested or Moluccan Cockatoo *Cacatua moluccensis*
Umbrella or White Cockatoo *Cacatua alba*
Philippine or Red-vented Cockatoo *Cacatua haematuropygia*
Tanimbar or Goffin's Cockatoo *Cacatua goffini*
Ducorp's Cockatoo *Cacatua ducorpsiis*

In this book I will discuss the six Australian species only, omitting the Cockatiel as there is a specific book available in this series, **A Guide to Cockatiels and their Mutations**, covering all aspects of this, the smallest cockatoo.

Although some of these species show sexual dimorphism (di=two; morphic=form/type), this can be minimal and they can still be very difficult to sex. Another feature of these cockatoos is their strong, powerful beak, varying in length, width and shape depending on the species, their diet and habitat. These birds display a crest either short or long, that is raised or lowered, usually when excited or alarmed, and have short, broad tails. All young are born with fine, yellow down, except for the Galah, which is born with a fine, pinkish down, and the Major Mitchell's Cockatoo being almost naked.

The white cockatoo family is synonymous with Australia. Apart from the Major Mitchell's Cockatoo, they are widely kept throughout this country as pets, rather than being kept and bred as aviary birds.

Most Australian white cockatoos, with the exception of the Eastern Long-billed and Western Long-billed Corellas, are plentiful in their respective native wild habitats, in fact, in some areas they are declared pests by farmers, with licences being issued annually by respective authorities to poison or shoot certain numbers. Perhaps, as they are so common in the wild throughout this country, aviculturists tend to keep and breed rarer bird species, leaving the white cockatoos for the pet market.

In saying this, however, I do feel that aviculturists in Australia are beginning to identify the different subspecies of the white cockatoo family, encouraging more interest and intrigue toward keeping and breeding this lovely group of birds.

MANAGEMENT

Purchasing Stock

There are several points to consider before purchasing any of the Australian white cockatoos either as pets or breeding birds.

View as many collections as possible to obtain as much information and ideas as you can. Contact local avicultural clubs, attend meetings and talk to experienced members. This can be done without commitment to joining a club until you are sure that you want to continue with bird keeping. Read and research the species of cockatoo you wish to keep, by reading avicultural books and watching videos, which can add valuable information.

Remember, cockatoos are much larger and stronger than most parrots or other birds, so thought must be given as to whether you are competent and confident in handling these birds, as you will probably have to catch them up at some stage of their life.

These species usually live for many years. Some individuals have been recorded to live between 80 and 100 years of age, so thought once again should be given as to who will care for or where these birds will be located if they outlive you.

As they are usually very noisy and screech loudly when alarmed, consideration must be given to neighbours if living in suburban areas and to any relevant council regulations. In certain states of Australia wildlife licences are required to own and keep some of these white cockatoo species, which incur annual costs as well as bookkeeping and periodical inspections carried out by the respective authorities.

Housing, food receptacles and water containers have to be made heavier and stronger than for other bird species, such as parrots and finches, which could prove more expensive.

Time and attention is also very important, as some birds can become prone to behavioural problems such as feather plucking. Experts cannot determine why some of these cockatoos begin this practice, although many believe boredom or stress could be responsible, (see *Feather Plucking*, page 67).

The best time to purchase stock in Australia is during the summer months (December - February), as the breeding season is usually over and there are plenty of young birds available.

You should never be in a hurry to buy the first bird or pair of birds you see. This applies to the white cockatoo species as well as other bird species. Look at them and make sure that they are healthy and in good feather condition.

Other questions to ask include
- How old are the birds?
- Have they ever bred?
- Are they related?
- Are the parents of the birds able to be viewed?
- What is their diet?
- Under what conditions have they been housed and managed?

Such information will be extremely important once you have your bird or birds at home and settled into their aviary.

It is always wise to purchase unrelated pairs for breeding. I prefer to purchase unrelated young birds and pair them together, waiting two or three years for them to breed. My reason for this is that once the birds have acquired their adult plumage, it is impossible to tell whether the birds are two or 20 years old, and you could be buying someone else's problems, eg non-breeders or even past their breeding age, mutilators or several other reasons why the owner may want to part with them. If you visit an aviculturist to buy birds and are told the ones that you are interested in are unrelated and there appears to be only one breeding pair in his/her aviaries - think twice before purchasing them.

Physical Examination

Never purchase a bird without first carrying out a thorough physical examination! When you have selected your cockatoo, ask for the bird to be caught up for you so you can perform your own examination.

Assessment of body condition can be done by feeling the amount of breast muscle coverage over the keel bone. Ideally, the muscle should curve convexly to come to a point at the sternum (breastbone). Birds in poor body condition will have a prominent keel bone giving a tent-like appearance. Emaciated birds will have very little breast muscle present. In cockatoos, obesity is a big problem. In severe cases, a 'channel' forms at the keel between the oversized breast muscle tissue which surrounds it on either side. Often there will be much fat both under and within the skin giving it a yellow appearance.

You should open out the wings, checking the underside of the flight feathers, looking along the central vein where lice may congregate. You can also ensure that the bird's wings are not clipped or damaged, through being kept in a small holding cage. White cockatoos produce powder down - a white, dusty feather by-product which coats the feathers, beak and feet, and your hands and clothes after handling! This has the effect of dampening the bird's colours eg Sulphur-crested Cockatoos have grey feet and beaks and bright, white plumage; Galahs have soft pink and grey feathering. Beware of white cockatoos with 'dirty' plumage, particularly if they have shiny beaks or poor feathering as this may indicate Psittacine Circovirus Disease (PCD)/Psittacine Beak and Feather Disease (PBFD).

Evidence of mites should also be checked. Mites are spread by direct contact and can be irritating to your bird and it will therefore rub the affected part of the body onto perches, which could cause a hazard to other birds in your collection.

Next, examine the eyes. The eyes should be fully open, bright, clear and free of discharge. The nostrils and cere should be clear with no matting of the feathers above them.

The beak should be smooth and preferably free of cracks, grooves or bleeding. The upper and lower mandible should meet correctly and should not be distorted. In most cases an overgrown beak can be clipped and reshaped back by an experienced aviculturist or avian veterinarian, however do take care that this is not an inherited weakness, as birds such as these should be avoided for breeding purposes.

Examine the bird's feet and toes carefully, as any damaged or missing toes could result in infertility in your breeding program. Overgrown claws can be clipped back with nail clippers. Slipped claws and leg damage should also be checked. The feet should be evenly covered in scales. Watch for wart-like growths which can sometimes be found on the feet of cockatoos. The undersurface of a cockatoo's foot will naturally be rough and uneven, due to the presence of numerous small bumps or papillae. When the underside becomes smooth and pink, the potential for infection and pain to develop are high. This condition is commonly known as bumblefoot. The causes are usually due to poor perching, poor nutrition, obesity and trauma.

The vent on any bird should be clean. Vent staining is often a sign that all is not well with the bird.

After careful examination of the bird, it is then your decision whether you purchase and take this bird home. Once you have taken possession of your new bird it is often very difficult to return it if you later realise that there is a problem.

It is certainly worth considering having a prospective purchase checked by an avian veterinarian.

If your cockatoo passes all inspections and you do decide to purchase it, no matter how fit and healthy it may look, do not overlook the importance of quarantine.

Transportation

Carry boxes are an essential piece of equipment for the

Carry box. Sturdy construction is vital when transporting cockatoos.

This method of freighting live birds being light in weight minimises air freight charges.

aviculturist, not only for transporting birds long distances by road, air or other means, but even more importantly for transporting birds within your own aviaries.

I have seen more escapees within an owner's property, whilst moving birds from aviary to aviary either by hand or catching net, than while transporting birds over long distances by road or air freight.

Carry boxes should be designed with minimal room inside, according to the size of bird you are transporting, to restrict movement and prevent injuries to a frightened bird.

Adequate ventilation should be provided, by designing the box with either an open wire front or ventilation holes in the sides.

Carry boxes for the cockatoo species must be of strong construction, as their ability to chew will soon have them out of the box and flying around in your vehicle, if this is neglected.

When designing or purchasing a carry box for a cockatoo it is a good idea to have it so that both ends of the box can be opened. This will make it easier to release the bird once you have it home.

If sending birds by air freight, a carry box can be constructed of strong or heavy wire and placed inside a ventilated cardboard box. This lightweight construction is acceptable to the airlines within Australia and will easily restrain the birds during the journey. Being light in weight will also minimise air freight charges.

Label with 'LIVE BIRDS' and indicate with arrows to the top of the box. Include the address, phone numbers and import/export licences of the sender and the receiver.

There are some basic rules that should be adhered to when travelling with or sending birds interstate.

- Birds should be transported in the passenger compartment of your vehicle and not the boot. This is law in some states of Australia, however, commonsense should tell us that birds could be susceptible to petrol fumes and lack of air causing them to suffocate.
- Only transport one bird per box. Even devoted breeding pairs of birds have been known, when under stress, to attack, kill or maim their partner.
- During summer, do not transport birds in the heat of the day, as this can cause stress or heatstroke.
- Provide seed and a piece of fruit, such as apple, for moisture.
- Try to minimise the time your cockatoo will be captive within the carry box.

As soon as you have released your birds, clean and disinfect the carry box. A simple way of making your carry box easier to clean is before each use place paper towels or newspaper, which is easily discarded, on the floor.

Keep birds separated when transporting. Note wire divisions in this carry box.

Quarantine

Once you have purchased new birds, they should not be released into their permanent aviary immediately, as all new arrivals should be quarantined. This is the practice of housing newly purchased birds **separately** from existing birds in a collection so that their health can be monitored. Ideally it will prevent the new bird from introducing any diseases which it may carry into the collection. It will also give new birds a chance to acclimatise to their new surroundings and management.

The type of quarantine facility can vary from the most simple to the most expensive. All that is required is a quiet, comfortable enclosure which is situated well away from other birds and their feed, furnishings etc, of a minimum dimension of 120cm cubed, in a warm, dry location. It should allow easy observation and maintenance of the birds.

A suspended cage located in a quiet room, back shed or garage is ideal. Suspended cages allow droppings to pass through the cage floor and so minimise the likelihood of birds reinfecting themselves via their droppings. Newspaper can be placed under the cage to catch faeces and urine samples and if necessary, this paper can be folded, placed in a plastic bag and taken to your veterinarian for analysis if there are any problems.

When setting up a quarantine cage, or even locating newly acquired birds in their permanent aviary, food and water receptacles should be placed close to a perch within easy reach of the birds. Newly acquired cockatoos can sometimes be reluctant, in a new environment, to fly to food and water containers placed on the floor or away from perches.

Above: Pet cockatoo cage can be utilised as a quarantine cage.

Aviary fittings, such as food containers, perches, or toys should be only used in the quarantine area and never transported or placed in any other cage or aviary. In order to prevent disease transmission, all dishes and equipment used in the quarantine area should be washed separately to those from the rest of the collection and disinfected. Quarantine birds should be treated last of all in the daily routine. Avoid traffic back and forth between the quarantine area and the existing collection.

How long should quarantine be? There is no exact answer to this question.

Unfortunately, some diseases may have very long incubation periods (ie the time from initial infection until the bird shows actual signs of the disease) so detecting these may take up to years.

However, in the majority of cases, a period of six weeks will allow many of the more common diseases to be detected, if present.

The quarantine should not finish just because the six week prescribed time limit is over, if there is any sign of disease, or if any test results are in doubt. If a bird becomes ill during quarantine, it should be treated and quarantine restarted from day one as soon as the bird appears healthy again.

On the day the new birds are to be released into their new aviary it is advisable to do so early in the morning, as this allows the birds all day to explore their new home, find seed, water and roosting places etc. Try not to spend too much time near the aviary on the first day, allowing the birds time to settle in. Observe from a distance.

Most importantly, make sure that the day you release your cockatoos is warm and sunny, especially if they have been quarantined inside, away from the elements for some time.

It is imperative that as soon as your new bird has finished quarantine and you have moved it into its permanent cage or aviary, that the quarantine cage, seed and water containers, perches and all other fittings within the quarantine cage are washed, scrubbed and disinfected, ready for future use.

Catching Nets

When selecting a suitable catching net for this species consideration should be given to the following.

- Size - diameter and depth of the net being sufficient for the size of species and to prevent escape.
- Padded rim - preventing death or injury when catching birds.
- Material - a closely woven fabric is preferable to a mesh fabric in which the bird may grip or claw and become entangled. A dark coloured fabric is preferable to a lighter colour.

Aim to catch the cockatoo in flight as this is the safest way to prevent injury to the bird. Avoid catching a bird during the heat of the day as this can increase stress to the point of death.

A catching net is designed to catch a bird and should never be used to carry or transport it, even from aviary to aviary.

It is very important to wash your catching net on a regular basis. They are easily washed and disinfected by hand without causing any damage to the net.

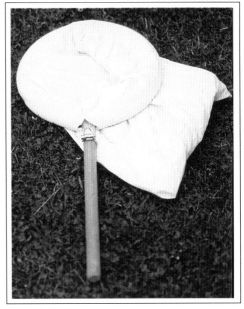

Above: Closely woven fabric is preferable to a mesh fabric.

Restraint

Everyone who owns and keeps birds will at some stage have to catch and handle them and it is very important that this is done with the minimum of fuss and stress for both bird and keeper.

Some aviculturists prefer to use gloves when handling cockatoos, however, I consider this method of restraint to be dangerous, as the gloves, although protecting the handler's hands, may make it very difficult to feel the pressure that you may be exerting on the bird. This pressure could cause death or serious injury to your cockatoo if care is not taken.

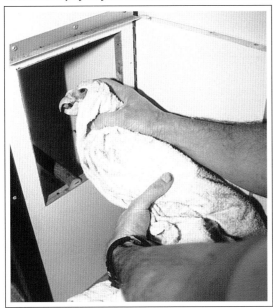

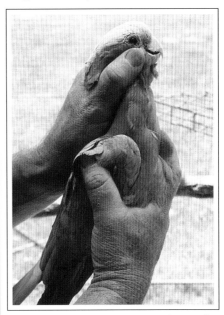

Above: Restraining your cockatoo for your own safety as well as the birds.
Left: Using a towel to restrain your cockatoo.

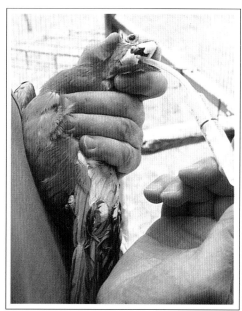

Left: Holding a cockatoo with one hand, so you can examine or medicate the bird.

Holding a bird too tightly around the chest area can restrict its air intake, causing suffocation. I suggest that bare hands give much more control, especially when handling such a large bird.

Hold the bird with one hand, using the thumb and index finger to grip the back of the neck area, restricting movement of the head, and making it impossible for the bird to move around or bite. Your other hand can be used to hold the feet and wings while you examine the bird.

Another method of restricting the bird's movements, while administering medicine or making physical examinations, is by holding the bird in the back of the neck area, as described previously, wrap its body in a towel or cloth, thus leaving a hand free to crop needle or examine the bird.

Observation

Observation is one of the most important rules of aviculture if you wish to have success. Today, most aviaries are set up to supply enough food to last a week or more, however, it is a wise practice to observe your birds at least twice a day, early in the morning and again before dark. This will alert you, before it is too late, to any bird that may have a problem.

During the breeding season or on excessively hot days observation checks on your collection should be carried out on a more frequent basis.

I often wonder why people keep birds and purport to being an aviculturist if they do not take the time to sit, watch and enjoy their collections. Surely this is what bird keeping is all about.

Many aviculturists are installing and using observation monitors throughout

Above: Observe your birds daily.

their aviaries to keep a close watch on their collections. Observation can bring to the fore signs of aggression, dissatisfaction with nesting sites or partners, neighbouring birds and diet, or signs of them being unwell.

Rodents

A strict eye should also be kept around the interior of your aviaries for signs of rodents such as rats and mice. Rats and mice can be a real problem for the bird keeper or aviculturist if they are not kept under control. These rodents leave many tell-tale signs in and around the aviary providing evidence of their presence.

* *Foul Odour*
  This odour can be evident in and around feeders, water containers and other fittings in the aviary.

These areas should be inspected, swept and kept clean at all times to discourage these vermin from congregating and becoming settled.

- *Droppings*
Wherever you detect the foul odour of mice you can be certain to see faecal droppings. These droppings should never be left in the aviary and should be swept up as soon as they are detected.
- *Seed Husks*
Seed husks piled together in a corner of the aviary are also a sign that mice are present. If an aviary is large and open, rats and mice will gather seeds such as sunflower and eat them in a quiet corner of the aviary leaving a pile of discarded seed husks.
- *Nests*
When rats and mice are allowed to overrun an aviary it is not uncommon for them to settle to such an extent that they build a nest and raise their families in this area. **This should never be allowed to happen!** Rats and mice should be controlled long before this can occur.
- *Holes*
Mounds of dirt and holes in and around aviaries should be dealt with to discourage the permanent residence of rats and mice.

Above: Mice holes in aviary floor. Mice should never be allowed to take over like this.

Below: Evidence of mice using a corner of an aviary.

Although it can be impossible to prevent rodents from entering your aviary complexes or cages, with the help of bait stations and by keeping food and water above the ground and out of the reach of such vermin, it is possible to keep them under control. Bait stations are easily made by sealing one end of a piece of 100mm diameter x 30cm long PVC piping and placing a screw cap fitting on the other end, with a hole large enough for rodents to enter. The cap is unscrewed and poison placed in the tube, the cap replaced, and the bait station positioned in a dark, quiet corner of the aviary.

Other methods of discouraging rats and mice in and around the aviary or aviary complexes are
- Flooding and filling in vermin holes as soon as they appear.
- Keep the aviary area clean. Do not place rubbish or other objects permanently along the aviary wall as vermin will use them as cover, often digging holes and living in and around these areas.
- Sweep and remove any spilt or disused seed in the aviary. Food and water containers should be kept off the ground and out of the reach of such vermin. Catching trays to contain seed spillage from food dispensers help keep the aviary floor clean and discourage rats and mice taking up residence.

Left: Poison bait to kill and keep mice under control in an aviary.
Below: Bait station used to keep mice under control in an aviary.

Pets

If choosing a pet cockatoo, a young handraised bird would be the best choice. Handraised young are usually obtainable just after the breeding season, however, when purchasing a just weaned bird, it is wise to obtain a diet sheet from the breeder, so you have a guide as to feeding over the next few weeks, to ensure that changes to the diet are not too drastic.

If you do not want to purchase a young bird and train it yourself, you can sometimes purchase a cockatoo that has already been trained, tamed and is already talking. Most people believe that cock birds make the best pets, becoming quieter and better talkers than hens. This is not always true, as many of the Australian white cockatoo species hens have proved to be excellent pets and talkers.

Apart from the Major Mitchell's Cockatoo, all other species of Australian white cockatoos make excellent pets, especially when taken from the nest and handraised.

In fact, some of these species are synonymous with suburban and rural backyards throughout Australia. These cockatoos are a very intelligent species and will bond strongly with human company. Therefore, it is understandable that the more time and effort that you attend to your pet cockatoo, training, talking and teaching tricks, the more you will be rewarded with the companionship of your pet friend.

Although handraised Major Mitchell's will remain as quiet as the other species of Australian white cockatoos, they tend to become aggressive as they mature, even towards their owners, and having no fear of humans, can inflict injuries to adults and children if not watched carefully.

The vocal abilities of cockatoos can be incredible, building up an extensive vocabulary, by mimicking their owners during normal daily routine. These cockatoos can also be taught tricks such as waving good-bye, playing dead, even riding a toy bicycle or car.

When teaching your cockatoo to talk or perform a trick, keep in mind that one or two, ten minute sessions per day, are better than half hourly or hourly sessions. Cockatoos will react better to a series of short, concentrated sessions held on a regular basis. Be patient, move and speak slowly and be gently persistent when training your cockatoo to talk or

These handreared Galahs make excellent pets.

perform a trick.

Pet cockatoos allowed to roam free in the house should **always** be under supervision. They are inquisitive by nature and can quickly get themselves into trouble with their habit of chewing most things with which they come into contact. Many household items can be toxic to pet birds, from products containing lead and zinc such as leadlight windows, old paint, curtain weights, jewellery, even some bird toys, to air fresheners and carpet cleaners, indoor plants etc.

Fumes from overheated teflon-coated non-stick frying pans and other cooking utensils can be lethal to birds also. Avoid potential sources of injury such as unshaded windows, naked flames from stoves and fireplaces and heated cooking utensils. In fact, it is best to keep pet birds out of the kitchen altogether to avoid potential disasters from occurring. Open toilets (yes, birds have drowned in these!), ceiling fans and open doorways or windows can all lead to mishaps.

Birds let out of their cage outdoors should again, **always** be under supervision. Birds not wing-clipped may be startled by a noise or predator, panic and fly away, no matter how tame they are. Wing-clipped birds cannot escape predators. Use commonsense and keep them supervised!

Anyone wishing to learn more about pet or companion birds would be wise to purchase a specialist book, such as **A Guide to Pet and Companion Birds**, outlining the keeping, training and well-being of such pets. There are also some excellent training videos available produced by world renowned bird trainer, Steve Martin, such as **Parrot Care and Training**.

HOUSING

Cages for Pet Cockatoos

When purchasing a cage for a pet bird as strong and as large as a cockatoo keep in mind that a cage can be too small but never too big. Cockatoos require a strong, roomy cage so they can move around and exercise as much as possible.

The minimum cage size recommended for a pet cockatoo is 90cm square x 90cm high, allowing the bird to extend its wings fully without touching any of the sides of the cage. Horizontal measurements, I believe, are more important than vertical in cage sizes, as birds tend to move from side to side rather than up and down.

There are a wide variety of pet cockatoo cages available, suited to both exterior and interior locations. The white cockatoo species within Australia have always been popular as pets and no doubt this has been realised by manufacturers opting to include specialised pet cockatoo cages as part of their product range. Choosing a cage for your pet cocky should not be taken lightly as there are important points to consider.

- Ensure that the cage you purchase is adequate in size and strength to accommodate your cockatoo.
- It must be of steel construction.
- Check the bar spacings are not wide enough for your bird to put its head through, leading to injury or death.
- It is important that the cage you purchase has a door and catch strong enough that the captive cockatoo cannot open.
- Some cages are designed with sliding trays underneath, to catch droppings, spilt seed and food pieces. These trays make cleaning and management of your birds much easier.
- Make sure the paint or coating on the cage will not flake off and is not harmful or toxic to your

cockatoo. A powder-coated finish is preferable.

- Check for tags or flecks of galvanising (zinc) which are toxic if picked off and eaten.
- If you purchase a stand for your cockatoo cage to sit on, make sure it is not only strong, but sturdy, especially when your pet is excited and flapping around.
- If you purchase a used cage, wash thoroughly and disinfect with a bleach before setting it up and introducing your cockatoo.

The advantage of many pet cages is that they are portable and easily transferred from one location to another. On a sunny, warm day they can be placed outside, hanging from a shady tree and brought back inside at night. There are also large, cockatoo suitable, patio cages which can be wheeled about from room to room or outside.

It should be stressed that wherever you place your pet cockatoo cage, make sure that it is not accessible to draught or direct sun.

In situations where cockatoos are kept in small cages as pets, owners sometimes allow their birds out of their cage to freely fly around inside their house to exercise and relieve boredom. A T-stand with seed and water

Above: T-stands are ideal for pet cockatoos.

containers attached, allows the cockatoo more time out of its cage and somewhere to perch. It must be stressed that at all times when a bird is out of its cage, it must be supervised.

Some people tether their pet cockatoos with a leg chain. It is my opinion that these types of restraints are totally unsuitable, and cruel for use with any bird as they can suffer from leg or internal injuries if they are startled, apart from being illegal in most states of Australia.

*Above:
Patio cage for housing a pet cockatoo inside your home or on your patio.*

*Right:
It is very important that your pet cannot put his/her head through the bars of the cage.*

Above: Backyard cage for a pet cockatoo.

Above left: Pet cages are available in a variety of designs.
Above right: Cage for a pet cockatoo constructed under the patio of the owner's house.
Right: Cage for a pet cockatoo.

Aviary Design and Construction

It is very important when designing an aviary or aviary complex that you not only consider the design of the aviary to accommodate your cockatoos, but thought also be given to the efficient feeding, cleaning and servicing of the collection. The easier this is made, the more pleasure and time you will have to enjoy your birds.

Over the years I have seen a large array of aviaries for housing the cockatoo species. The designs of some of these aviaries vary greatly, some superior to others. So, how do you design and construct your own cockatoo aviary?

Firstly, don't be in a hurry. Investigate all local government by-laws to ensure your plans to build an aviary meet the necessary requirements. (These vary from area to area). Give thought as to the location of your aviary. It is a delight to look out of your lounge or kitchen window directly into your aviaries in the backyard, however this may not always be practical.

Aviaries, where possible, should face north - north-east (in the southern hemisphere) as this will provide maximum daily sunlight and protection from inclement weather, including prevailing southerly wind and rain.

If you have no other choice but to face your aviaries south or west, thought must be given to protecting the aviary front, preventing the birds from being exposed to bad weather.

Another important factor in positioning your aviary is not to have them facing your driveway where your birds will be exposed to car lights at night.

When planning the design of the aviary, consider the cockatoo species you would like to keep and how you would like to keep them. Some people house their birds in single pairs, while others prefer a colony aviary housing the same species. This will determine the aviary dimensions, as the more cockatoos you plan to keep in one aviary, the larger the construction requirements. Personally, I do not recommend colony aviaries as too many problems seem to arise from this style of housing, such as fighting, breeding stress and behavioural problems.

Other bird keepers choose to build individual aviaries around their backyards, set into gardens to give a natural display effect.

Cockatoo aviaries that are specifically set up for breeding are usually built in blocks or complexes which are divided into a number of separate flights. The entrance doors of each flight then lead into a walkway. Walkways are essential to minimise escapes. These walkways can be attached to either the rear or front of the aviaries. Other benefits of walkways are that the entrance doors to each flight can be made to a full height making entry much easier, instead of having to bend over to enter these flights. Plan the walkway to be wide enough to enable access with a wheelbarrow for cleaning purposes. A recommended minimum width of 1.5 metres will also allow you to utilise this space as a small bird room, for storing seed, medicines or any other avicultural supplies.

Self closing doors fitted with a spring hinge are also essential to minimise escapes.

When housing cockatoo species, timber structures are not viable as these birds will destroy such construction in no time at all. Aviaries for cockatoos should be constructed of steel and metal which is very reasonably priced and easily worked with.

The dividing walls between breeding pairs in my aviaries are solid iron sheeting as opposed to wire as I feel that this prevents fighting and offers more privacy between pairs, particularly during the breeding season. If wire partitions are used the divisions should be double wired between pairs to eliminate fighting and injuries caused by neighbouring birds.

Cockatoos are hardy birds and enjoy being exposed to the elements (sun, rain etc.) and it is for this reason that aviculturists housing these birds prefer to leave at least half to three-quarters of the roof area open to the weather, however, this will depend greatly on the area in which you live and the weather conditions.

The amount of money available for you to spend on your new aviary will also have to come into planning consideration. Plan your aviary with the option to extend at a later date, if finances are limited. Bargains in the form of second-hand aviaries are available from avicultural club members' sales or through magazine or newspaper classified advertisements. One of the advantages of purchasing an aviary or aviary complex second-hand is that you can view the aviary, fully erected, at the owner's property and consider the suitability for your future plans. It is very important to remember that any second-hand aviary must be thoroughly scrubbed and disinfected prior to releasing your birds into it.

Aviaries and aviary complexes have improved in design and construction over the past 10 to15 years. Manufacturers are designing lighter, higher quality materials which are just as effective as the heavier steel used previously. This makes the purchase of such materials much less expensive. There is a large array of colorbond roofing iron available in panel rib, trimdeck, custom orb or other profiles and in a variety of colours. This will eliminate painting, making your aviary or aviary complexes blend naturally and inconspicuously into your landscaping.

There are basically two aviary designs used by aviculturists, conventional and suspended.

Conventional Aviaries

The recommended minimum conventional aviary size to house one pair of Australian cockatoos is 4 metres long x 1.5 metres wide x 2.2 metres high. If room on your property permits, the larger the better.

Conventional aviaries are erected from ground level, thus allowing the birds access to the ground or floor. This style of aviary design is very popular with aviculturists in Australia and can

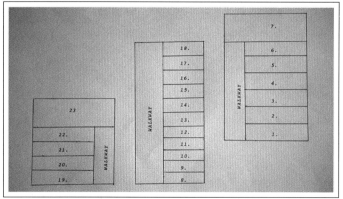

be landscaped to resemble the birds' habitat, using logs, ponds, natural perches and grasses to give them a natural look. However, cockatoos will destroy plant foliage and almost anything that is attempted to be grown in their aviaries.

Being built from the

Left: Draw up a plan of your aviaries before construction.

ground up, conventional designs need to be rodent, or in some areas snake proof, which can be difficult as these pests do not need a very large hole to enter the aviaries. To keep such vermin from entering, concrete or brick footings can be erected and sunk below ground level to a minimum depth of 50cm.

Flat iron or other suitable sheet materials may also be used for this purpose. On my aviaries, flat iron continues to

Above: Walkway at the rear of an aviary complex.
Left: Design walkways to enable you to manoeuver a wheelbarrow and carry out maintenance to your birds easily.

a height of 1.2 metres above the ground, which also gives the cockatoos security from animals such as dogs, cats or in rural areas, foxes that may wander around the perimeter.

Conventional aviaries with earthen floors, do give a natural look. However, paved or concrete floors are easier to keep clean and more hygienic for your cockatoos. They are very easily serviced by simply sweeping or if thought is given at the slab stage of construction, natural fall can be built into the floor so that they can be hosed down and the water drained away. Concrete floors will also help to combat pests such as snakes, mice and rats from entering your aviaries and if they do gain access they are much easier to eradicate.

Suspended Aviaries

Suspended aviaries are very popular as they offer many advantages. However, I do not feel that all the white cockatoo species benefit from this aviary design.

As the name suggests, suspended cages or aviaries have a wire or weldmesh floor situated 1 to 1.2 metres above ground.

The advantages of suspended cages are that once birds are wormed and introduced to this

Above: Suspended aviaries.

housing the likelihood of them being reinfected is minimal, and pests and vermin have difficulty entering these aviaries.

There are many ways of constructing suspended aviaries or cages. They can be constructed as all wire cages and enclosed totally within a shed or birdroom, or built as a row of cages with a walkway built at the rear, with the roof of the walkway extending to cover at least half the roof area.

Left:
Large, well-built cockatoo aviaries to house a large collection.

Right:
This large aviary is suitable for colony breeding cockatoo species.

Left:
Small backyard cockatoo aviary. Note small, open, attached cage for the occupant to gain access to the elements.

Above: Suspended aviaries,
Right: Aviary set-up suitable for breeding a single pair of cockatoos.

Consideration should be given to providing privacy to individual pairs in the form of solid partitions between cages.

The recommended minimum size suspended aviary for Australian white cockatoos is 1.2 metres wide x 1.2 metres high x 4 metres long.

My reason for not using this type of aviary complex is that the Australian white cockatoo family like to forage and fossick on the aviary floor, which of course is impossible with this type of housing.

Above: High fronts on conventional aviaries offer your birds more security from dogs, foxes and cats.
Left: Cockatoo aviary complex. Position aviaries north - north-east.
Below: Large cockatoo complex.

Wire Size and Gauge

The wire you choose for your conventional or suspended aviary needs to be of a heavy gauge to accommodate the powerful beaks of these cockatoos.

For the smaller species such as Galahs and Short-billed Corellas, 16 gauge wire should be sufficient to restrain them within their

aviaries. There are three wire aperture sizes in 16 gauge that I would recommend, these being 12.5mm x 12.5mm, 12.5mm x 25mm, and 25mm x 25mm, although smaller wire sizes do make viewing your cockatoos more difficult.

For the larger species such as Sulphur-crested Cockatoos, Long-billed Corellas (Eastern and Western) and Major Mitchell's Cockatoos, 16-12 gauge wire mesh 12.5mm x 25mm or 25mm x 25mm square is recommended. I consider any wire greater in aperture than 25mm x 25mm square would allow small wild birds, snakes and vermin to enter the aviary, contaminating the seed and water, and being a potential threat.

Due to the galvanising process of wire during manufacture, it is advisable to wash or scrub any newly purchased wire with a water and vinegar solution and remove any excess galvanised tags or flecks. Remember also to pick up any metal particles formed when drilling holes for screws to attach wire etc. These galvanised particles can cause heavy metal poisoning to your cockatoos which, if undetected and if immediate medical attention is not sought, can be fatal. Remember, that after the wire wash with water and vinegar, thoroughly hose down the wire. BHP Wire Product's new Evencoat™ wire has largely eliminated the galvanised particles on the wire, making it safer to use with birds.

Left: Woven wire is not suitable to use on cockatoo aviaries.
Below: Double wire partitions are recommended between breeding pairs of cockatoos.

Woven or netting bird wire is totally unsuitable for use on cockatoo aviaries as these birds have been known to undo this wire, which can lead to escapes.

Many aviculturists paint the wire of their aviaries with a black or dark coloured acrylic paint, which can serve two purposes. Firstly, it will help prevent and protect the birds from heavy metal poisoning and secondly, the birds within the aviary will be easier to view from the outside.

When constructing an aviary complex with a view to keeping smaller cockatoos such as the Galah and Short-billed Corella, it is wise to build the aviary using heavier materials and wire than required for these species. If you decide in the future to purchase larger cockatoos, these aviaries will accommodate them and a new aviary does not have to be constructed.

Perches

Perches made of natural tree branches are recommended as the variation in thickness will provide exercise for the cockatoos' feet. Perches for the cockatoo species can present a problem as their chewing habits can at times be unbelievable. For this reason I prefer tea-tree or red gum branches ranging in diameter from 5cm-8cm. It does not seem to matter how thick or strong timber perches are, cockatoo owners always seem to have the never ending task of replacing them.

The method of fixing or changing perches within the aviary must be made as simple as possible. I fix a perch holder that is made from a heavy gauge flat iron, folded to the shape of a V, either side of the aviary wall. The perch is simply measured to length

Above: Painting the wire on aviaries enables you to view the occupants more easily.

and placed in the bracket. This method is very simple but also very effective, as a perch can be replaced in seconds.

I know of other aviculturists who use 4cm x 5cm red gum timber which is purchased from their local timber and hardware store. These are simply fixed and easily replaced by using pergola brackets that are attached to the aviary wall. For large cockatoo complexes a bundle of 4cm x 5cm red gum timber can be purchased, stored and when perches are required, cut to length and placed in the aviary.

I have met aviculturists who are frustrated with the task of replacing timber perches and instead used metal water pipe or tubing. I do not recommend this as metal materials can become very cold and slippery during winter months, and can lead to foot problems.

Perches should be fastened securely, as any movement of the perch will cause the bird to feel insecure and become stressed. Unsecured perches can also lead to infertility problems during the breeding season as when the cockatoos are balancing on a moving perch, proper copulation may be difficult.

A minimum of at least two perches should be provided in the cockatoo aviary - one at the front of the aviary where the birds will have access to sunshine and rain and another at the rear of the aviary under shelter, to offer the birds security and protection from the weather.

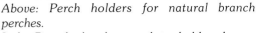

Above: Perch holders for natural branch perches.
Left: Pergola bracket used to hold red gum perches.

Cockatoos, being relatively large birds, need room when they fly, so it is advised to allow a clear flying space and not clutter the aviary with perches, as this may lead to injuries such as broken bones from the birds clipping these objects during flight.

Habitat

In the wild, the Australian white cockatoo species are distributed over a wide range of habitats including rainforest, eucalyptus bush, open forests, mallee and farmland areas.

When providing accommodation for cockatoos it is very difficult, due to their destructive nature, to provide a natural environment or habitat within their aviary. Natural perches and logs are really all that can be provided, as any natural growing plants or trees would soon be demolished. I provide metal or PVC pipes that are fixed to the aviary

Above: PVC pipes fixed to the aviary wall to hold fresh tree branches.

wall and in which I regularly place tree branches. Although these branches do not last very long they do look good and help prevent boredom. Before you cut any perches or branches from trees to furnish your aviary, it is imperative that you check beforehand that they have not been chemically sprayed or are not poisonous to your birds.

Left: Due to the cockatoo species destructive nature, it is not practical to provide natural habitat within their aviaries, however fresh branches can be provided.

Food and Water Containers

Stainless steel containers are recommended for seed, food and water, as they are easily cleaned and disinfected.

Fruit, vegetables or green food placed in the aviary should be located where it will not become susceptible to contamination ie not under perches. Foods should not be thrown or placed on the aviary floor.

I provide a feeding tray on which I place apple, corn, suitable wholegrain bread, carrots, endives, milk thistle, seeding grasses and any other tidbits, including nuts from flowering native trees or bushes.

Food trays are inexpensive and very easily constructed from 15mm square steel tubing with heavy gauge 25mm x 25mm wire, tack welded to the bottom of the frame. 25mm angle iron is then welded to the underside of the frame, which allows a stainless steel colorbond tray to be slid in underneath the wire to catch any discarded food. Catching the discarded food helps to keep the aviary floor clean and therefore prevents the cockatoos from eating any used or contaminated food. Each day spoiled food can be slid out and the tray emptied, washed, sterilised and replaced.

Water bowls, as with all other cockatoo utensils, need to be constructed of heavy materials in order to prevent your birds from tipping them over. If ceramic, they should be glazed, as this will not only help to prevent algae growth, but makes scrubbing and cleaning water bowls much easier. Terracotta or galvanised water containers are unsuitable as they make any in-water medication (drugs, wormers, vitamins, calcium etc.) impossible. These types of containers absorb the medication from the water.

Above:
Note feeding station on doors to allow easy servicing and feeding.

Right:
Dry seed and soaked seed mix in stainless steel bowls.

Above: Automatic watering system.   Above: Stainless steel water bowl.
Below: Revolving stainless steel feeding bowls.

When positioning water containers or bowls in your cockatoo cage or aviary, do not place them under perches or roosting places, to prevent fouling. Locate in a shaded area of the aviary, away from direct sunlight, as in extremely hot weather the water in these bowls can become unbearably hot and the growth of bacteria increases rapidly. Water can also be used in sprinkler systems throughout your aviaries. Generally, the Australian white cockatoo species do not bathe, although if a sprinkler system is installed in their aviary, these birds will take full advantage of the situation. It should be noted that any exposed pipe ends of watering or sprinkler systems that are run throughout your aviaries must be metal piping or tubing and not plastic.

Night Lights

Night lights should be considered in your design. Night lights are inexpensive, easy to install and the running costs are minimal. They are not a bright light, giving out only a soft glow so that within the aviary it is still dark but offers enough light to enable a frightened bird or birds to find their way back to the safety of their nest, perch or roosting site.

It is recommended that lights are set up and installed in the walkway area. This is beneficial as the light from the walkway, normally located at the rear of the aviary, entices the bird or birds towards the rear if frightened, therefore the birds do not spend the night in the cold and are not accessible to predators as could be the case if the birds are hanging on the front wire all night. Also, lights placed in the walkway prevent birds from chewing the lights or wires connecting the lighting system, which could be fatal and even dangerous to the aviculturist if using a 240 volt system.

Another advantage of night lights is that parents have more hours of light to feed their young in the nest. The low glow of night lights enable you to see anything in or around the aviary area, such as dogs, cats, possums and rodents, that may disturb your birds at night.

Security

Security is really only prevention and cannot be guaranteed 100% burglar-proof. However, it will usually make a thief think twice before attempting to steal from you.

Many larger aviary complexes are fenced off with a high cyclone wire fence which cannot be guaranteed to prevent a theft, but will make access more difficult. Some breeders run two or three

Left: Guard dogs are always a deterrent to potential thieves.
Below left: Security cameras can also be used to prevent thefts. Typical video surveillance set up.
Below right: Reliable and strong padlocks are the most important precaution of all.

dogs within these enclosures to alert them to any unusual activity. Dogs used for this purpose are usually purchased as young puppies so the birds get used to them and they don't upset the collection.

Electric fences are also a good security device making an unsuspecting burglar unsure of what to touch as well as keeping predators away from your collection.

Smaller backyard collections can be made secure by having the yard fenced off and locked, making access difficult.

Electric or 'magic' eyes are also a very good investment and are relatively inexpensive. They create an invisible electronic beam which, when broken, will sound off a receiver, that can be carried on your person whilst going about your daily chores or kept inside your house at night. The magic eye will alert you of visitors long before your doorbell rings.

Security cameras can also be used to minimise thefts. These may be set up in and around aviary complexes with a TV monitor in the house. Many aviculturists install security camera surveillance systems connected to sirens or horns to alert them of intruders and deter stealing.

One of the most important security tips is to make sure that you do not become so caught up in installing security such as magic eyes, surveillance cameras and other systems that you forget the most important precaution of all - padlock your aviaries at all times with a reliable and strong padlock.

FEEDING & NUTRITION

Some cockatoo species can be prone to obesity and while it is important that you provide a nutritious, well-balanced feeding program for your captive birds, a strict eye on diet should be maintained. Obesity can lead to many problems in captive cockatoos, not only causing bad health, but also breeding difficulties.

Carbohydrates, fats and proteins are the primary sources of energy for the bird, supplying the fuel necessary for body heat and work.

Carbohydrates

Carbohydrates are the chief source of energy for all body functions and muscular exertion and are necessary to assist in the digestion and assimilation of other foods.

Fats

Fats or lipids are the most concentrated source of energy in the diet. When oxidised, fats furnish more than twice the number of calories per gram furnished by carbohydrates or proteins. In addition to providing energy, fat acts as a carrier for fat soluble Vitamins A, D, E and K. By aiding in the absorption of Vitamin D, fats help make calcium available to body tissues, particularly to the bones and beak.

Protein

Next to water, protein is the most plentiful substance in the body. Protein is one of the most important elements for the maintenance of good health and vitality and is of primary importance in the growth and development of all body tissues. It is the major source of building material for muscles, blood, skin, feathers, keratin and internal organs, including the heart and the brain.

Protein also helps prevent the blood and tissues from becoming too acid or too alkaline and helps regulate the body's water balance.

Foods containing protein may or may not contain all essential amino acids. Vegetables and fruit are incomplete protein foods. To obtain a complete protein meal from incomplete proteins, one must combine foods carefully so that those low in essential amino acids will be balanced by those adequate in the same amino acid.

Vitamins

Natural vitamins are organic food substances found only in living things, that is, plants and animals. Each of these vitamins is present in varying quantities in specific foods and each is absolutely necessary for proper growth and maintenance of health.

Dry Seed

A dry seed mix is supplied to my cockatoos at all times throughout the year. The selection of seeds, vitamins and minerals for the cockatoo species is the vital difference between excellent birds and average birds. If your cockatoos are fed a varied diet and you know the benefits of the seed or food you are feeding them, they should be acquiring all the vitamins and minerals needed for healthy birds.

Some people tend to think that the larger the bird or cockatoo the larger the seed to feed them. This is not always so. Cockatoos need a well-balanced diet just as we do and it is for this reason that you should not just select a seed because it is cheap or pre-packed and sitting in the shop to be purchased, but you must also understand the health benefits of each of the individual seeds contained in a mix.

A reputable seed merchant will explain the nutritional benefits of the seeds you require, and supply and mix the specific ingredients.

Above: Solid containers for feeding dry seed within the aviary.

I feed my cockatoos a mixture of Budgerigar seed with a small amount of added grey-striped sunflower seed. This Budgerigar mix is comprised of 33% Japanese millet, 10% canary seed, 10% panicum, 32% French white millet and 15% hulled oats. Once all these seeds are mixed thoroughly the mixture is coated with a Vitamin A and E rich oil (not cod liver oil).

It is very important that the dry seed mix is not placed in the aviary where it could be subject to becoming damp or wet, as this can lead to the transmission of disease and the deterioration of the seed quality.

The storage of dry seed is also very important. All seeds should be stored in ventilated containers. The original packaging, designed to allow breathing of the seed is usually ideal.

Left: Storage containers for keeping seed dry and safe from vermin.

The transfer of seed to airtight containers should be avoided unless a desiccant (drying) agent is included. All seeds contain between 5% and 15% moisture and will sweat if isolated in an airtight container. This will lead to germination of the seed and development of fungus and bacteria.

Soaked and Sprouted Seeds

When seeds are immersed in water the process of germination begins. The sprouting process increases the level of easily digested sugars (maybe the reason why birds relish soaked or sprouted seed) and the level of vitamins and amino acids essential to life.

A soaked seed mix consisting of equal parts of mung bean, wheat, barley and grey striped sunflower is given daily throughout the year - this mixture being allowed to sprout during the breeding season.

Above: Pine cones can be given to cockatoos to relieve boredom.

Prepare by soaking the seed mix in water and a cleansing agent such as Aviclens™ for 8-12 hours, rinse thoroughly with fresh water, drain and feed to your birds. If you wish to sprout these seeds, leave for a further 24 hours after draining, until the seeds start to germinate.

A frozen vegetable mix containing peas, carrots, corn and beans is thawed and mixed with the soaked or sprouted seeds.

It is stressed that soaked or sprouted seeds should not be left for long periods in the aviary, as bacteria may develop. It is also recommended to use stainless steel bowls which are very easily cleaned and disinfected to prevent bacteria related diseases.

Fruit and Vegetables

It is important to feed a variety of food types other than seeds in order to maintain a balanced diet. Fruit, vegetables and green food are all essential food items. Although these items have a very high water content, they contain essential fibre and vitamins which are not available in the dry seed mix.

Above:
Dog biscuits, almonds and peanuts are readily accepted by the cockatoo species.

Left:
Fresh vegetables, bread and corn frozen in readiness for the breeding season.

Cockatoos, as with most other birds, will devour a variety of fruits, such as apples, oranges, grapes, peaches, pears, watermelon, passionfruit and rockmelons.

Vegetables and green foods can be purchased in frozen packs, thawed and fed or mixed with soaked or sprouted seeds, as previously discussed. Relished fresh vegetables include corn on the cob, silverbeet, endive, broccoli, carrots, peas, celery and capsicum. When in season, I purchase some 2000 corn cobs from a local vegetable market gardener, which are frozen, this supply lasting for approximately 12 months. Daily requirements are thawed.

Seeding Grasses and Weeds

Seeding grasses and weeds, such as naturally grown and freshly picked (semi-ripe or ripe) millet, milk thistle, New Zealand spinach, dock, panic, veldt and rye grasses, are common when in season and are thoroughly enjoyed by cockatoos. When collecting seeding grasses or other tidbits from paddocks, roadways or vacant allotments you must be sure that they have not been sprayed, as this could be fatal to your birds.

It is for this reason that all seeding grasses given to my birds are home-grown in a bird proof enclosure. During certain times of the year seeds such as millet, canary, sunflower, oats and barley are sown in this enclosure providing ripe and ready to feed seeds all year.

Advantages of home-grown seeding grasses inlcude

- Free from chemical sprays.
- Ensures a plentiful supply of seeding grass throughout the year.
- Is convenient and time saving.
- Reduces your green food and seed expenditure.

Right: Bird-proof enclosure in which home grown seeding grasses can be grown.

Bottlebrush
Callistemon salignus

Supplementary Foods

Supplementary foods, to provide extra protein, vitamins, minerals and in some cases as therapy to prevent boredom, can also be fed to your cockatoos as daily food sources.

Cockatoos will readily accept peanuts, walnuts, brazil nuts, almonds, bones such as cooked chicken, lamb or pork chops, provided these are removed before they spoil. Various types of wholegrain breads, boiled eggs and dog biscuits will also be enjoyed.

The seeds from various trees can be collected weekly, kept in containers and fed daily. Tree branches or bushes will provide many hours of entertainment and nutrients for cockatoos, climbing and chewing nuts and leaves. This activity can assist in prolonging the life of perches and prevents boredom, which can lead to psychological problems such as feather plucking.

Examples of some of our native eucalyptus gum nuts that can be gathered and given to your cockatoos are illustrated in the accompanying drawings.

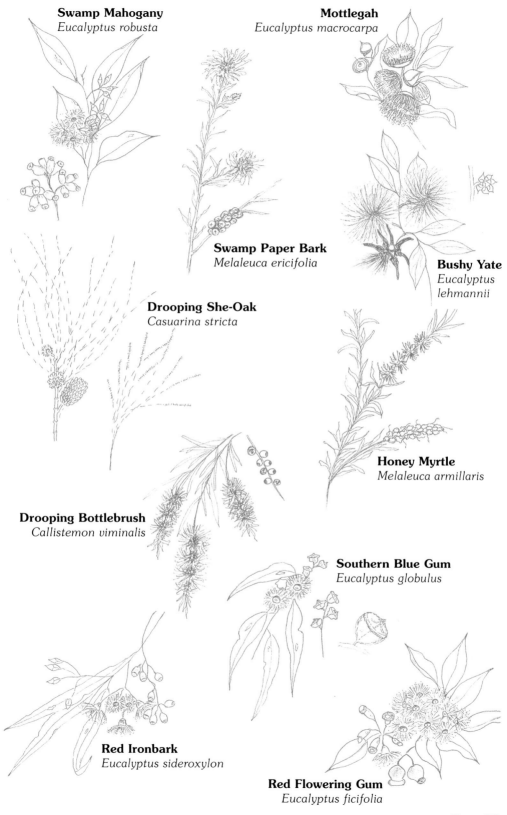

Swamp Mahogany
Eucalyptus robusta

Mottlegah
Eucalyptus macrocarpa

Swamp Paper Bark
Melaleuca ericifolia

Bushy Yate
Eucalyptus lehmannii

Drooping She-Oak
Casuarina stricta

Honey Myrtle
Melaleuca armillaris

Drooping Bottlebrush
Callistemon viminalis

Southern Blue Gum
Eucalyptus globulus

Red Ironbark
Eucalyptus sideroxylon

Red Flowering Gum
Eucalyptus ficifolia

Pelleted Food

Some aviculturists prefer to feed only a pelleted diet, as a balanced nutritional food source. Commercially formulated pellets can be provided as a supplementary food, particularly during the breeding season. However, I believe that it is vital to supply a variety in size, shape, texture and colour of food items to these species to prevent boredom with diet. Formulated diets should therefore be supplemented with vegetables etc, as described earlier.

Grit and Calcium

Grit and calcium are essential to the cockatoos' diet, providing beneficial components for the birds' skeletal structure and the formation of the egg shell.

They should be available at all times of the year, however, hens consume grit and calcium products on a more regular basis during the breeding season.

I prepare a grit mix by combining medium shell grit, charcoal, cuttlefish bone and washed fine river sand. These ingredients are ground and mixed thoroughly together using an old meat mincer, which can be purchased at a very reasonable price from a Sunday market or garage sale.

Calcium blocks, beneficial to the birds, are also home-made. The following recipe is one I have successfully used.

Above: Grit and calcium mix.
Below: Old meat mincer used to grind up the calcium and grit mix which is then fed to the cockatoos.

Ingredients
1 25kg bag casting plaster (Plaster of Paris)
1 kg sulphur
1 kg medium shell grit
1 kg calcium carbonate
1 kg seaweed meal
5mls iodine (liquid) mixed in 16 litres of water.

Mix the dry ingredients and add the water/iodine mix to form a paste-like substance. Place into two-litre ice-cream containers (an ideal size) and when set remove for use.

It is worth mentioning that cuttlefish bone and calcium blocks can often be insufficient sources of calcium to cockatoos due to them not ingesting enough. In water calcium supplementation may be required before and during the breeding season.

Left: Cuttlefish bone.

Water

Water constitutes over 50% of a bird's total body weight. As with humans, they lose water through evaporation and excretion in the urine, which is greater than through normal metabolism.

Fresh, clean water should be provided daily.

Many water authorities adulterate the water supplies, adding chemicals such as chlorine, fluoride and other agents, as well as scouring out water mains, stirring up bacteria and dirt, which eventually flushes out of our house taps. I prefer to provide rain or tank water. Catch water from the aviary roof (which would normally run onto the ground) in a small 250 gallon rainwater tank which is then piped into each individual aviary.

Above: Fresh water provided daily is essential.
Left: Rain water tank to supply fresh water within the aviary.
Below: Automatic watering system.

Converting Birds onto a New Diet

It can be difficult to convince some birds to try new dietary items. Many older birds can be very set in their ways. The following are some suggestions which may help to achieve this. No matter which technique is used, the bird should be monitored for signs of distress or insufficient feeding to sustain itself. Under no circumstances should a bird be starved. Avoid changing the diet at times of stress, such as immediately after purchase, during illness, moulting or breeding. However, perseverance and sometimes ingenuity may be required for a successful outcome, such as

- Limit the time during which seed is available. Seed may be offered for 10 minutes in the morning and then again in the evening only. At other times, all seed should be removed and the new foods provided. This gives the bird enough food to survive but keeps it hungry enough to encourage it to try the new foods. This is particularly useful for complete diet changes, such as from seeds to pellets.

Below: Cockatoos also like to be showered with a fine spray of water.

- Birds should be started on a nutritious and varied diet from a young age when they are more inquisitive and accepting of new food items. Weaning is the ideal time, but most birds are purchased as independent youngsters or adults.
- Try mixing the new diet with the old. Seeds can be embedded into fruits or vegetables or pellets can be moistened, squeezed into a ball and have the seeds mixed through them.
- Feed new items in the morning when the bird is most hungry.
- Use a 'teacher bird' which is already eating the desired diet and keep it within visible distance of the bird which is to be converted.

BREEDING

Establishing Breeding Pairs

If breeding success is to be achieved, compatibility of your cockatoo pairs is very important.

It is recommended to purchase young, unrelated cockatoos and pair them up, even if it means waiting three or four years for them to breed. Due to the long life span of these cockatoo species it is advisable to be patient and buy young birds, as once you have compatible breeding pairs you should have many years of chick rearing.

Some breeders purchase several unrelated hens and cocks of the same species, housing them together in a large aviary and after observing, eventually select the birds that pair up naturally for their breeding program.

With the exception of the Major Mitchell's Cockatoo, this cockatoo group can be colony bred as individual species (preventing hybridisation). It should be noted that any colony housing of cockatoos should be monitored carefully, especially during the breeding season, and any troublesome birds removed immediately. It is very important that you introduce all the birds into the aviary at the same time (this is also advisable when pairing and housing individual breeding pairs).

Pairing birds for breeding, whether housed as single pairs or in a colony system, can be a very dangerous time of a bird's life. In a colony, bullies can persecute and stress individual birds to the point of death. There is often one dominant pair of birds in a colony system and although this pair may breed and produce young on an annual basis, the other pairs within the colony may not be as successful. Observation will alert you of imminent problems.

I prefer to house my breeding pairs of cockatoos individually, as this offers the birds more privacy and less distractions which undoubtedly leads to better breeding results.

When pairing cockatoos for the purpose of breeding, provide a choice of two or three nesting sites (logs or nestboxes). Once the cockatoos have chosen a nesting site the other logs or boxes can be removed from the aviary.

In the wild, Australian white cockatoos usually lay a single clutch (although commencement of breeding season and clutch sizes vary between species). However, in captivity, if eggs are taken for artificial incubation or the young are taken from the nest at an early age for handrearing, it can be possible for your breeding pairs to double brood.

As most of the young I rear are taken for handrearing (for the pet market), my breeding pairs double brood, therefore calcium is increased throughout the breeding season. This is done by providing liquid calcium supplement, such as Calcivet™ or Calcium Sandoz™, in the water and extra cuttlefish bone and grit in the aviary.

Healthy cockatoos normally have a natural instinct to breed. Possible reasons for lack of breeding success are

- The cockatoos are not a true pair (ie cock and hen).
- The cockatoos are incompatible (you may have to change or swap your pairs around).
- The hen may not be satisfied with the nesting facilities (try hanging a variety of logs in the aviary to give the hen a choice).
- The cockatoos may not be in breeding condition (they may be obese and therefore the diet may have to be reviewed).
- Lack of privacy. The birds may be too busy fighting or worrying about neighbouring birds in the aviary (they may have to be moved to alternate accommodation or solid partitions built between neighbouring pairs).

- A bird may have a physical problem, such as feet or leg deformities and may not be able to balance properly during mating.
- Birds may be sick or in poor health and therefore unable to breed (veterinary consultation may be required).
- Cockatoos are immature (you may have to wait three or four years).

Hybridisation can occur between any of the Australian white cockatoo species and it is for this reason that different species and even subspecies should never be housed together.

I feel as responsible bird keepers or aviculturists, we should discourage interspecies pairings or breedings, endeavouring only to breed pure and true species (mutations excepted).

Following are some breeding statistics of the Australian white cockatoo species.

COCKATOO SPECIES	No of Eggs Per Clutch	Approx Egg Size	Egg Laying Interval (days)	Incubation (days)	Fledging Period (weeks)
Sulphur-crested Cockatoo	2-3	38mm x 28mm	2-3	25-28	9-13
Short-billed Corella	2-4	41mm x 29mm	2-3	26	7-8
Eastern Long-billed Corella	2-3	51mm x 30mm	2-3	24	7
Western Long-billed Corella	1-3	51mm x 30mm	2-4	26-28	6-8
Major Mitchell's Cockatoo	2-3	38mm x 28mm	2-3	24-26	7-8
Galah	3-4	35mm x 26mm	2-3	23-25	6-7

A common feature of all these species is that incubation is shared by both the cock and hen. Cocks generally sit during daylight hours with the hen relieving him and sitting during the night.

Due to the long life and breeding cycles of these species, owners do not need to replenish their breeding stock with new birds, as proven breeding pairs will generally produce chicks for many years.

Short-billed Corella eggs.

Short-billed Corella chick - 1 day old.

Short-billed Corella chicks - 1 week of age.

Short-billed Corella chicks - 3 weeks of age.

Short-billed Corella chicks - 5 weeks of age.

Short-billed Corella chicks - 10 weeks of age.

Nestboxes and Logs

In the wild, Australian white cockatoos will accept a variety of nesting sites, including some that we may even think to be unusual. When we consider wild nesting sites, hollow logs in trees immediately come to mind, however, there have been many unusual sites recorded. An example are the cliffs along the mighty Murray River between Murbko and Blanchetown, South Australia where cockatoos (mainly Sulphur-crested Cockatoos)

Above: Wild Sulphur-crested Cockatoos nesting in the cliff faces along the Murray River.

roost and nest in the cracks and crevices of the clay cliff faces, between 20 and 90 metres above the river.

The urge to breed can be so strong that if normal nesting sites cannot be found in the wild, these birds will naturally adapt and take advantage of any potential nesting situation available. Therefore, I believe they will also adapt to a variety of nesting facilities in aviculture, although hollow logs are preferable. Natural logs can be very heavy and awkward to position in the aviary, however the advantages should be considered, including

- They are more natural for the cockatoo species to breed in.
- Cockatoos like to chew and work the log during the breeding season. By using natural logs this is made possible.
- Logs are more insulated and do not overheat during hot weather temperatures as do other nesting alternatives, such as man-made metal nesting receptacles.
- Humidity within a natural log is more stable.

Above:
Hollow log with an open top hung on a slight angle.

Left:
Ex-army gunpowder barrel with natural spout.

Right:
Metal rubbish tin used as a nesting receptacle. Lid has been removed for cleaning.

Due to the difficulties that some aviculturists have in obtaining hollow logs, wooden nestboxes can be constructed, as an alternative. Cockatoos will go to nest in boxes quite readily, however due to their destructive natures these will normally only last a couple of breeding seasons before having to be replaced.

During construction some aviculturists place a short piece of hollow log on top of the nestbox to encourage the cockatoos to enter. The dimensions of the nestbox I use are 30cm square x 60cm to 90cm deep. A wire ladder is placed on the inside of the box to allow easy access in and out for the parents. This ladder helps to prevent eggs or chicks being damaged by the parents, especially if they are startled or frightened.

Metal nestboxes may also be provided in the form of rubbish tins or even ex-army gunpowder barrels. The advantages of these nesting receptacles are that they are usually readily obtainable, lightweight, easily installed and easy to fumigate, keeping them disease free.

A natural log or spout can be placed on the entrance to entice the cockatoos to enter. As metal nests can become very hot during the summer months (usually when your cockatoos will have young), it is essential that they are not placed in direct sunlight or in a hot area of the aviary. It may even pay dividends to provide ventilation.

Above: Untreated pine nestbox with natural spout used for breeding Galahs.
Below: Removable bottom in nesting receptacles make cleaning and fumigating of the nest at the end of the breeding season much easier.

If you are providing logs or nestboxes in a colony breeding aviary, it is wise to provide at least two nesting receptacles per pair of birds housed. They should be hung at the same height allowing as much distance or space between nesting sites as possible. This is suggested to keep fighting over nesting sites between pairs to a minimum.

After each breeding season all nesting logs, boxes or metal nesting containers should be emptied, cleaned and fumigated in readiness for the following breeding season.

My nesting logs have a detachable bottom to allow easy cleaning and fumigation. At the end of each breeding season I remove the bottom of the log to empty the old nesting material. A gas blowtorch is then lit and placed under the open bottom, burning out and killing any bugs, mites, lice or diseases that may be lurking or breeding in the log.

Once the bottom has been replaced, a mixture of 50% sawdust and 50% peatmoss is used for nesting material which is laid 10cm to 12cm deep in the bottom.

I also spray around the inside and outside of my aviaries and nesting receptacles with Coopex™ residual insecticide (made up as per manufacturer's directions) to kill insect pests such as ants, beetles, cockroaches, silverfish, fleas and spiders. Pests such as these could be the reason that a particular hen refuses to enter a nest, or abandons eggs or young in the nest.

Sometimes I feel that we are too quick or eager to blame the parent birds for nesting failures, rather than our own lack of observation and preventative measures.

From experience, I have found inspection openings in any nesting receptacle are most important, as breeding hens and young need to be inspected daily. Chicks may also need to have leg rings placed on them and such openings will make access much easier.

Regular nest inspections are essential. The young chicks should be removed from the nest with periodical checks carried out on them. A check on the chick's eyes, nostrils and beak should be made to ensure that they are free from any nesting debris which may be caught in their eyes or blocking their nostrils or beak. The young bird's feet should also be cleaned to make sure that no nesting material or faeces are caught or built up between the bird's toes or leg ring.

Record Keeping

In aviculture, record keeping is vital and combines well with the observation of your cockatoos.

I keep a daily diary on hand and as I do my aviary rounds each morning, record any unusual or different happenings within the aviaries. These records can be consulted as references or if any problems arise with a bird.

If a cockatoo is sick, records should be kept of the problem, such as
- How long has it been sick?
- What action has been taken?
- How long has the bird been in a heat source?
- What medication are you using?
- How long is it on medication?
- Did the bird die or survive?

Breeding records should include
- Which pair of cockatoos are the parents?
- How many eggs and when were they laid?
- When did the young hatch?
- How many hatched?
- When were they leg rung or microchipped? What are the ring numbers or microchip details?
- How many fledged?
- What sex are the young?
- Who were the young sold to?

Incubation and handrearing records should include
- Incubator and brooder temperatures and observations.
- Formula details.
- Feeding volumes and intervals.
- Weight and growth rate records.
- General handrearing observations.
- Physical development of chick.
- Weaning foods and details.

There is an old saying, 'we should learn by our mistakes'. Unfortunately, if we do not keep records, occurances are sometimes easily forgotten and not only do we fail to learn but others will not learn either.

There are many simple methods of keeping records. They may be kept in the form of a diary, card system, note book, breeding register or computer programs, designed for this purpose. Record books may be purchased from bird shops or avicultural clubs.

Identification

To enable you to keep accurate records of your cockatoos, you must be able to identify each bird individually. There are two recognised methods of achieving this, leg ringing or the placing of a microchip under the bird's skin.

Leg Ringing

Leg ringing (banding) chicks is an easy and convenient method of identifying your birds.

There is a large selection of leg rings available, ranging in colour, size and number combinations to suit all birds. Rings can be personalised, indicating the breeder's initials and numbers on the rings.

There are two types of leg rings available - split rings which can be opened up and placed around the leg of any cockatoo whether young or adult, and closed rings which must be placed on the cockatoo's leg between the age of 10-14 days, although some species need to be banded earlier. I use stainless steel, numbered, closed rings on my cockatoos' legs for identification and have never found them to be a problem.

I do not recommend split rings for the cockatoo species as their powerful beaks could crush the ring closed tightly on the leg, resulting in severe damage or in some instances the loss of the leg.

The correct size of the leg ring for the cockatoo species that you are ringing is very important. If it is too tight, damage to the leg can occur and if too loose, the cockatoo can catch the ring and become entangled in the aviary, resulting in injury or death.

When closed ringing a cockatoo chick, hold the three longest toes forward, slipping the ring over these three toes. Push the ring up over the joint of the foot, sliding the smaller toe back between the leg and the leg ring. The back toe can be flicked out using a toothpick or small, pointed tool between the leg and the toe to pull the toe out from under the band.

Above: Leg rings are available in a range of colour, size and number combination.

SPECIES	RING SIZE	APPROXIMATE AGE OF RINGING (Days)
Sulphur-crested Cockatoo	13mm	10-14
Short-billed Corella	10mm	10-14
Eastern Long-billed Corella	12mm	10-14
Western Long-billed Corella	12mm	10-14
Major Mitchell's Cockatoo	10mm	10-14
Galah	10mm	10-14

*Leg ring sizes are a measurement of the inside diameter·

The approximate age of ringing is a guide only and a strict eye should be kept on the growth of chicks, as some birds develop much faster than others.

 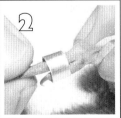  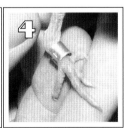

Series of photographs showing fitting of leg band.
1 and 2: The band or ring is placed over the three longest toes and moved down until it meets the fourth toe.
3 and 4: Gently pull the toe through the ring until it is located correctly on the chick's leg.

Microchipping

Microchips are small electronic devices (silicon chips) that are injected into birds and other animals to provide permanent identification. These chips are coded with a series of numbers and letters that can be read by a transponder (scanner). Microchips are implanted by a veterinarian, generally by injection, into the pectoral muscle of the cockatoo. The chip code is recorded at the time of implantation and is registered (for a further fee) with a central registry on computer. Microchips can therefore be used as proof of identification and possibly assist in deterring theft of your cockatoos.

Right: Microchipping equipment. From top to bottom - applicator, identification tag, needle and silicon microchip.
Below:
Microchip transponder.

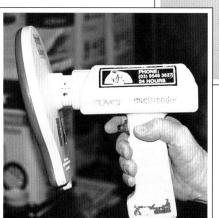

Incubation

The practice of incubation and handrearing of birds is not as difficult as it once was, although it does demand a great deal of time and commitment, such as having to be prepared to get out of bed to check eggs or feed your baby cockatoos and ensure that their crops are emptying.

Anyone who is serious about incubating and handrearing cockatoos should be prepared to research and study this practice, before attempting it themselves. In this section I will give a brief outline as to how I incubate and handrear my baby cockatoos.

Although I handrear Sulphur-crested Cockatoos, corellas and Galahs for the pet market, I only incubate eggs under an extreme emergency, preferring chicks to be two or three weeks old, incubated and fed by the parents, before removing them for handrearing.

Incubation from the egg is a long and demanding task that not always ends rewardingly.

When contemplating incubating and handraising cockatoos, hygiene, temperature, humidity and diet are the four basic requirements for success. When the eggs to be incubated are taken from the nest they are cleaned, marked with a pencil for identification purposes, dated and recorded in a record book so that the development of the embryo in the egg can be monitored.

The incubator dry temperature is set at 37.6°C (99.7°F) and the humidity or wet bulb thermometer reading set at 27.75°C (82°F). The automatic turning device within the incubator turns the eggs 180 degrees every 60 minutes.

Fertility and embryo development is monitored and recorded by candling the egg, ie using a light source. Candling is also required to check eggs close to hatching, as the egg should not be turned three or four days prior to hatching.

Once the egg begins to pip it is moved to an incubator set at 36.9°C (98.5°F) dry temperature and a higher humidity. The egg is not turned.

On hatching, the chick remains in the incubator for a further 24 hours before being transferred to the brooder. I use a Kimani™ Post Hatch Brooder manufactured in Western Australia. The brooder is thermostatically controlled and is extremely accurate and stable. Fan-forced heat circulates evenly without blowing directly on to the chick.

Chicks in the brooder have a totally safe environment, isolated from direct contact with the fan, heating element and water dishes. Brooder temperature over this time is decreased, depending on the comfort of the chicks and how quickly they develop their feathers. The chicks remain in the brooder until fully feathered.

If you watch carefully you will see that chicks that are too hot tend to be restless, having their mouths open and wings spread out. Chicks that are too cold will become less active, usually huddling together, and digestion will slow.

BROODING TEMPERATURES

DAY	TEMP °C	DAY	TEMP °C	DAY	TEMP °C
1	36.5	16	30	31	28
2	36	17	30	32	28
3	35.5	18	30	33	28
4	35	19	30	34	28
5	34	20	30	35	28
6	33	21	30	36	28
7	32.5	22	29	37	28
8	32	23	29	38	27(Room temp)
9	32	24	29	39	27
10	32	25	29	40	27
11	32	26	29	41	27
12	31.5	27	29	42	27
13	31	28	28	43	27
14	31	29	28	44	27
15	31	30	28	45	27

The temperature will vary according to the species, depending on how much natural down they have, whether they are a single chick, a group of chicks and how quickly they are developing.

Once fully feathered, when artificial heat is no longer required, the chicks are removed from the brooder and placed in a weaning box to entice them to eat alone and stabilise prior to locating into a weaning cage, then eventually an outside aviary. A weaning box can be made by using a plastic storage box and folding a piece of 12.5mm x 12.5mm bird wire to fit inside the box approximately 50mm above the base, to keep the chicks away from discarded food and faeces. A weaning cage fitted with low perches is recommended for the chicks to get used to perching and wing flapping exercises prior to release into the aviary.

After use, the brooder is thoroughly cleaned and disinfected.

Handrearing

When chicks are first hatched they may be dehydrated and weak. Before feeding the chicks handrearing formula they should be rehydrated by feeding an electrolyte such as Hartmanns™ solution and a nutritional source such as Ensure™. The electrolytes rehydrate and provide body salts. When rehydrated (approximately 48 hours), a thin formula is fed to the chicks. The formula is gradually increased in volume as the chicks require, until the feed intervals and emptying process of the crop are balanced.

APPROXIMATE FEEDING INTERVALS OF AUSTRALIAN WHITE COCKATOO CHICKS

Age (Weeks)	Interval (Hours)
1-2	every 2-2.5
3-4	every 3
5-6	every 5
7-8	every 7
9-10	every 12

Note: These intervals are approximate only. The chick's crop should be checked to ensure that it is emptying correctly prior to each feed.

Commercial handrearing mixes available are undoubtedly the best we could have, as veterinarians, chemists and nutritionists have worked together to produce such products, giving your young birds the best possible start in life.

Personally, I prefer to use Lakes™ handrearing mixture, as I have never had any problems while using this formula. Other recommended and popular commercial handrearing mixes available include Roudybush™, Pretty Bird™, Vetafarm™ and Wombaroo™, which are all beneficial to your young cockatoos. It is most important to follow the manufacturer's mixing directions to obtain maximum success.

The prepared handrearing mix is fed to the chick at a 41°C to 43°C thermometer reading. During feeding, the temperature of the mix is re-tested by placing a spoonful of formula to the inside of the wrist. Formula temperature is very important, as if it is too hot not only can this affect the contents of the formula, but may also burn or scald the chick's crop, especially if food is microwaved. If the formula is fed too cold it could chill the chick or cause crop related problems.

There are several feeding instruments available, including the syringe, crop tube, crop needle and bent teaspoon. Some handrearers prefer to use a crop needle, stating that the volume of food given to each chick can be measured exactly in the syringe and that the task of handrearing is made quicker and cleaner. I prefer to use a spoon (with the sides bent up) although it can be a slower feeding process. I find the volume of food for each chick can still be measured in a feeding cup and if the birds are cleaned well after each feed, using a damp face washer or baby wipe, they will be weaned just as clean as a crop needle or crop tube fed bird.

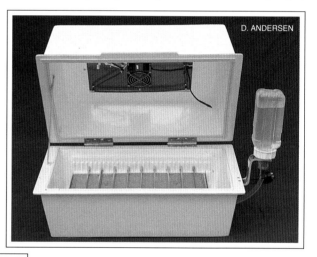

All cups, spoons, crop needles, crop tubes or syringes used should be thoroughly washed and disinfected in Miltons™ antibacterial solution (or similar) after each feed.

If you have more than one chick to handrear at the same time, separate feeding utensils should be used for each chick. This is recommended to prevent disease or infection transmission between birds. Some aviculturists use plastic disposable spoons (bending the sides up), which are disposed of after each feed.

Above: Kimani™ incubator used by author.
Left: Kimani™ Post Hatch brooder used by author.

Not enough can be emphasised regarding the cleanliness and sterilisation practices when incubating and handrearing. All equipment including brooders, feeding instruments, chicks, and especially, your own personal hygiene, must be kept spotlessly clean.

Weight records are vital when handrearing, as weight loss in a chick will probably be the first sign of a potential problem.

Weigh the chick daily, before its first feed, recording the weight in a book. These records not only give you an indication of the chick's progress, but will also provide a reference if you have a problem with a chick in the future.

AVERAGE WEIGHT CHART OF
AUSTRALIAN WHITE COCKATOO CHICKS

All weights are shown in grams.

Day	Sulphur-crested Cockatoo	Short-billed Corella	Eastern Long-billed Corella	Western Long-billed Corella	Major Mitchell's Cockatoo	Galah
15	566	264	205	116	102	99
16	588	281	205	130	120	112
17	595	270	204	143	134	126
18	602	273	220	161	146	148
19	624	298	237	177	160	157
20	641	305	254	184	178	171
21	651	314	271	199	188	186
22	669	338	286	209	208	197
23	680	360	303	228	228	212
24	701	354	319	250	242	226
25	703	349	330	255	260	224
26	713	367	348	273	266	241
27	715	376	375	289	284	249
28	737	372	381	307	290	262
29	750	380	385	337	296	266
30	771	384	408	341	308	274
31	777	385	413	369	318	280
32	780	380	421	374	320	275
33	775	380	440	390	320	262
34	785	379	444	401	320	255
35	788	375	447	412	328	259
36	790	375	451	422	336	263
37	800	373	447	444	340	267
38	798	371	452	464	342	271
39	795	350	453	476	348	262
40	792	346	457	487	345	263
41	807	357	468	492	348	267
42	795	357	478	497	347	260
43	798	359	483	501	347	265
44	803	357	492	517	345	268
45	821	350	492	528	342	262
46	811	351	496	553	343	265
47	811	349	497	558	345	265
48	812	350	493	574	345	269
49	809	349	485	582	346	266
50	809	Weaned	Weaned	599	346	271
51	810			Weaned	345	Weaned
52	809				346	
53	810				Weaned	
54	Weaned					

It should be noted that weights between the same species of cockatoos that you will handrear will vary considerably, depending on their progress in the beginning and the age they were taken for handrearing. These weights are average, as looking back through my records, some species have weaned a heavier and some a lighter weight.

The weight chart indicates the average progressive weight gains of the Australian white

cockatoo species that I have taken for handrearing at 15 days of age.

Looking at the chart you will see fluctuations in the weights of certain birds at particular times. This is due to the chicks beginning to pick at weaning food on some days. During this time chicks will not eat as much formula, preferring to feed themselves, as they are naturally preparing to fledge and fly. At this stage of their life, feeding intervals are increased to fewer feeds per day to encourage chicks to eat on their own.

Weaning dates will also vary between birds of the same species. Some birds will be naturally slower to wean than others or they may even be spoilt and demand food from you to seek attention. Don't be fooled by this, as in the wild, it is not uncommon for chicks to beg for food from their parents long after they are weaned and fully independent.

Above: Chicks in a weaning box with raised wire base to keep them away from discarded food and faeces.
Left: Don't ever leave chicks unattended.

Suitable weaning food includes corn, peas, sprouted seed, soaked seed, fresh fruits and grain bread.

It is very important that the crop and droppings of weaning chicks are checked, to be sure that the bird is ingesting food. Droppings should indicate a change in colour, substance with some bulk and the presence of urates separate to the urine and faeces. As the chick progresses on weaning foods, the droppings will develop more like that seen from an adult bird.

The weaning period can vary greatly. Two or three chicks wean quicker together than a single chick. I believe the competition and inquisitive natures of a group of cockatoos picking at the food is the reason that groups of young birds will wean quicker.

Right: Healthy five day old Sulphur-crested Cockatoo chicks.

Once they are eating alone and have been exercising their wings, they are placed in an outdoor holding aviary, where I monitor them closely for a further four to six weeks before selling.

In some states it is law that a bird cannot be sold until it is self-sufficient. I personally will not sell a handraised cockatoo until it is four to five months old ensuring that the new cockatoo owner will have no problems with the bird. Most people who purchase handraised cockatoos for a pet are not experienced bird keepers and should not be placed in the situation of 'finishing off' or looking after a cockatoo that is not fully weaned.

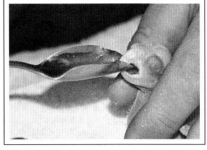

Above: Spoon feeding a young cockatoo,
Left: Nest of Galahs taken for handrearing.

There are some very good books and videos available on incubation and handrearing that I thoroughly recommend, including

- **A Guide to Incubation and Handraising Parrots** by Phil Digney, Published by **ABK Publications**.
- **Parrots - Handfeeding and Nursery Management** by Howard Voren and Rick Jordan.
- **Parrot Incubation Procedures** by Rick Jordan.

Above: Formula and tools needed for handrearing.
Right: Scales and record keeping - essential when handraising birds.

Sexing

If you are purchasing a pair of cockatoos for the purpose of breeding then it is essential that you choose the correct male and female gender. Birds can be identified as being monomorphic or dimorphic species. Monomorphic species show the hen and cock to look alike and dimorphic species exhibit visible distinguishing features.

Of the six Australian white cockatoo species discussed, the Galahs, Sulphur-crested and Major Mitchell's Cockatoos are dimorphic, although the variation in the degree of difference between true pairs can be minimal.

The Short-billed, Eastern and Western Long-billed Corellas are monomorphic and need to be surgically or DNA sexed to ascertain gender.

Above: An example of dimorphism by eye colour. Hen (left) has a lighter coloured iris than the cock.

Surgical Sexing

The surgical sexing procedure must be performed by an avian veterinarian and involves anaesthetising the bird.

The procedure is relatively quick, taking only a few minutes. In many instances the surgical incision is small enough so that sutures are not required.

The advantages of this technique are that the results are immediate and if the birds are closed rung or microchipped, then a sexing certificate signed by the veterinarian can be issued. Visualising the organs enables the veterinarian to determine how well developed the bird's sexual organs are, how likely it is to breed in the short and long term, and whether any abnormalities are present which may prevent the bird from breeding. Certainly with adult birds, the ovaries and testes are obvious.

The disadvantages of surgical sexing include the small risk associated with any surgical procedure.

These tend to be few and usually involve a pre-existing illness or when birds are sexed at too young an age.

DNA Sexing

DNA sexing involves the taking of body samples, either blood or feathers, from which the DNA is isolated and examined at a laboratory which can then differentiate cocks from hens.

The advantages of this safe method is that samples can be taken from birds without the need to anaesthetise. Birds don't have to be caught and transported to the veterinary clinic which minimises stress. In addition, DNA sexing allows the sexing of even the youngest chicks, which could not be surgically sexed at such a young age. Most laboratories offering this service provide sample collection kits, which can be mailed. This is helpful to aviculturists living in remote areas.

A consideration in DNA sexing is that the laboratory relies on the sender's honesty that the sample being tested has indeed come from the bird indicated by the sender. Therefore, unless the laboratory itself has taken the blood sample, or the sample was taken by a veterinarian or other unbiased third party, there can be no guarantee that a particular DNA sexing certificate applies to a particular bird.

DNA sexing gives no indication of the current health of the bird, its readiness to breed, or reproductive abnormalities which may preclude a bird from breeding in the future. There is also a small time delay from when the sample is taken to when results are obtained, (usually days).

The decision on which technique to use will depend on the particular circumstances and requirements of each individual aviculturist. Both techniques have helped to take the guesswork out of sexing cockatoos and have allowed the establishment of young pairs which are more likely to bond and therefore breed successfully.

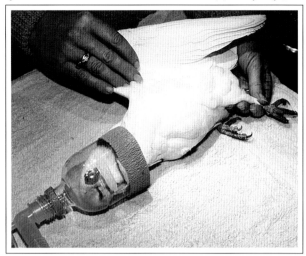

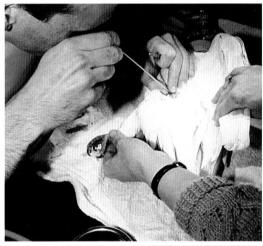

Above:
Anaesthetising a bird in preparation for surgical sexing.

Left:
Surgically sexing a Western Long-billed Corella.

Below:
Endoscope used for surgical sexing.

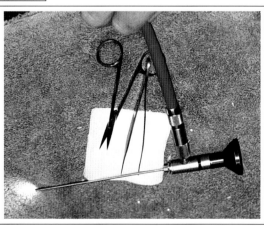

DISEASES AND DISORDERS
COMMON TO COCKATOOS

Recognising a Sick Bird

Birds are well-known for their ability to hide signs of disease until they are too ill to continue hiding their illness. This is referred to as a preservation response or reflex. A sick bird in a flock is

- Persecuted by its flock-mates and either driven out or even killed, as a result of the ongoing struggle for social status, or to stop it from drawing the attention of predators to the flock itself.
- More likely to be singled out by a predator and killed, especially by birds of prey.

It therefore follows that by the time a bird shows signs of being unwell, the disease process is well advanced. It is thus very important for the bird keeper to detect signs of illness as early as possible. The best way to do this is to know what is normal for your birds and note any variations from this. It is the person who spends time observing and therefore knowing his/her birds, who is most likely to do this successfully.

Some early changes are very subtle and difficult to describe, such as changes in a bird's perching posture, how keenly it eats or perhaps just changes in its behavioural routine.

The healthy bird should sit bright-eyed and upright on the perch with tight feathering. The wings should be carried evenly and the birds should use both legs with ease.

It is best to observe a bird from a secret vantage point where possible, to detect its true behaviour, as even a sick bird will tighten its feathers and look alert when approached.

From afar, healthy birds at rest will perch on one leg, however some like to remain on both legs for short periods, eg when dozing or during cool weather. In hot weather conditions, most cockatoos will perch on both legs, with their wings held out from the body. The trick is to know when this goes on for longer than is normal.

Watch for birds sitting fluffed for extended periods with the head under the wing, birds spending unusually long periods on the aviary floor, shivering, straining to pass droppings, tail bobbing or exaggerated breathing, coughing, sneezing, vomiting, excessive drinking etc.

If at all suspicious, it is best to catch the bird and perform a physical examination as described on page 9.

Examining a bird's droppings is an important way in which a bird's health can be assessed and can often indicate the presence of disease before the bird shows other more obvious signs. A healthy bird's droppings consists of three parts. The solid, dark coiled part in the middle is the faeces. This should be formed (ie tubular in shape) and may vary from dark green to black in colour. The white, paste-like portion is known as the urates, and the surrounding layer of clear liquid, the urine. The urine is usually noticed in fresh droppings, after which it is either absorbed or it evaporates. Changes in the amount, colour and consistency of these three components should be noted as they may indicate disease. If the bird needs to be taken to an avian veterinarian, it is best to leave paper on the bottom of the cage for the veterinarian to see the droppings. Collecting a fresh sample in gladwrap or foil will help to keep a dropping sample fresh until it can be taken to a vet.

Examining how much a bird is eating can also give warning to a bird's state of health. Although this sounds obvious, it is amazing how many birds that appear to eat, merely pick up the food in their beaks and then drop it. Taking the time to see that a bird is actually eating its food is very important.

What to Do with a Sick Bird

Once it has been decided that a bird is unwell, there are three courses of action available.
1 Do nothing and hope it improves.
2 Catch the bird and put it into a hospital cage.
3 Take it to an avian veterinarian.

Option 1

The wishful thinking approach. If you choose the first option, you are more than likely going to end up with a dead bird. A harsh statement, perhaps, but often true. As mentioned previously, by the time a bird is noticed to be ill, the disease is usually well advanced, so the bird is beyond recovering on its own.

Putting medication in the drinking water is often a futile effort as most ill birds won't eat or drink enough medicated water to effect a cure. In any case, one would have to guess the correct treatment for the particular bird's illness.

Option 2

The Hospital Cage. Three of the most important needs of an ill bird are heat, water (hydration) and an energy source. Birds spend much of their energy maintaining their body temperature at an optimum level. By providing the bird with heat, it can utilise its limited energy to fight the illness and maintain other body functions. Heat can be provided by a heating lamp or globe, or by heating a small room. Ideally a temperature of 26-30°C should be maintained day and night for a sick bird.

Regardless of the heat source, ensure that the bird cannot directly touch it and burn itself. Make sure that there is a bowl of water in the cage to add moisture to the air to help prevent the bird from becoming further dehydrated.

A hospital cage.

Do not forget to ensure that there is adequate ventilation whilst providing protection from draughts. Personally, I do not suggest totally enclosed, glass-fronted hospital cages as they provide poor ventilation and the air within quickly becomes stale. I have also noticed that glass fronts on hospital cages tend to frighten some birds.

The hospital cage should be situated in a quiet spot with subdued light to allow the bird to rest.

Hydrating the bird can be achieved by administering fluids orally. By adding an electrolyte and energy source to the water, one can help the bird to maintain its energy levels and further help recovery. Products such as Polyaid™ and Emeraid™ are specifically designed for this use and when mixed according to the manufacturer's instructions, can be given directly to the beak.

If these are unavailable, a solution of glucose water or honey water can be used in an emergency. This can be thickened with baby cereal such as Farex™ if the bird is underweight. Alternatively, handrearing formula appropriate for the species can be administered, preferably by a crop tube or needle.

Crop feeding a bird is a very useful skill and should be learned by all serious aviculturists. If you don't know how to do this, ask an avian veterinarian or fellow aviculturist to demonstrate to you the procedure.

Please note: in very weak birds, crop dosing can be dangerous as they may passively regurgitate food/water which may end up being breathed into the windpipe, air sacs or lungs, with fatal results. Such birds require immediate veterinary attention.

For cockatoos, between 5mls-10mls of fluid can be given at a time. It is better to give a small volume more often than to overfill the crop and risk the bird choking. Larger volumes can be administered later on.

A hospital cage is most useful for giving basic first aid to an ill bird, however if the bird has not recovered within a few days, veterinary attention will be required.

A basic first aid unit for bird keepers should include the following.
- Hospital cage.
- Crop needles, feeding tubes and syringes.
- Electrolyte and energy supplements.
- Oral calcium for egg binding and egg laying weakness eg Calcivet™ or Sandocal™.
- Topical antiseptic and wound dressing eg Betadine™ solution (never ointment/gel).
- Worming medication.

Do not, under any circumstances, use ointments on birds. These are most commonly used to treat eye problems or wounds on the body. The problem is that they mat the feathers, thereby decreasing the bird's insulating ability which can be disastrous for a sick bird.

The stickiness of ointments encourages food and faeces to adhere to the bird increasing the risk of infection and adding to the bird's discomfort. Birds preen and may ingest the ointment, which can be poisonous, further adding to its problems. In all circumstances liquid forms of topical medications are preferable.

Resist the temptation to dose the bird with an antibiotic. Antibiotics should only be used under veterinary supervision. Often, ill birds are brought to the veterinarian after being treated with 'over the counter' antibiotics by the owner, in a vain hope of correcting the problem. Not only are these antibiotics often ineffective for the bird's problem, but they may affect any tests that the veterinarian may need to perform.

In some cases they may make the problem worse, eg if the bird has a fungal problem. They can alter the normal bacteria in the bird's digestive tract, interfering with any immunity that these 'good' bacteria may provide. It is imperative to try and identify the cause of the illness so that the **correct** treatment can be given. An example of inappropriate use of medicines is giving antibiotics to vomiting chicks suffering from yeast infections.

Option 3

The decision on whether to take a sick bird to an avian veterinarian will depend on several factors including how ill the bird looks, the bird's value (and that of the collection) and the proximity and availability of an avian veterinarian. It should be remembered that the more time that a bird is delayed from receiving professional treatment, the sicker the bird is likely to get and the harder it will be to save the bird.

It is preferable that the person who actually cares for the bird takes it to the veterinarian. This allows the veterinarian to obtain important information regarding the caging, feeding, length of illness and other relevant questions which may support a diagnosis.

Pet birds kept in cages should be brought in their cage. Do not clean the cage before the visit to try and impress the vet! It is far more helpful to see the bird as it is normally kept as this may provide some useful information.

Aviary birds should be taken to the veterinarian in a covered carry cage or box, as this will minimise stress. If the box has a removable front panel, it will allow the vet to observe the bird before catching it. Ensure that the door to the carry cage is large enough to allow the removal of the bird whilst in the hand.

Paper should be placed on the bottom of the cage so that any fresh droppings passed can be taken for testing. It may be useful to bring fresh droppings passed at home wrapped in foil or gladwrap.

Any other items which the owner may consider important should also be brought, eg new foods, medications, shed feathers (for birds with feather problems) and photographs of aviaries etc.

When arranging an appointment with the veterinary clinic, it may be useful to ask the staff exactly what is needed to be brought in. Unless the veterinarian is a long way away, it is best to remove water dishes to prevent spillage or spoilage of droppings, or else place cotton wool soaked in water as a moisture source.

It is best to locate and become acquainted with your nearest avian veterinarian before an emergency arises. There is a list of avian veterinarians at the back of each issue of ***Australian Birdkeeper Magazine***. In some instances, an experienced avian vet may not be close to your area. In these circumstances it would be wise to find a local veterinarian who is at least interested and familiar with birds. Even if he/she cannot directly diagnose the problem, by contacting an experienced colleague a diagnosis and effective treatment can be instituted and the bird hopefully saved.

The Dead Bird

Unfortunately, every aviculturist will have to deal with the death of a bird at some stage in their lives, no matter how good an aviculturist one is.

The options as to what to do are again three in number.

- Firstly, the bird can be disposed of. Although certainly the cheapest option (in the immediate and short-term) it provides no information as to the cause of death. If an infectious or toxic disease is involved, further deaths may occur.
- The second option is to examine the dead bird yourself ie perform your own post-mortem examination (necropsy). Some diseases, such as roundworm or tapeworm infections may be obvious, while others cause very subtle changes which only an avian veterinarian or pathologist may be able to correctly interpret. In addition, there may be more than one disease process occurring and these may be missed by the untrained person.

The main problem with do-it-yourself post-mortem examinations is that well-meaning people have in the past incorrectly diagnosed causes of death because of their lack of training, and disseminated this misinformation as fact. The classic is the bird that died from trauma, ie 'a broken neck'. Just because a bruise can be seen under the skin overlying the skull does **not** mean that the bird died of trauma. Blood naturally pools in several areas after death and the skull is one of these. Unless the skull is actually damaged or bruising of the brain or spinal cord can be shown, then the cause of death lies elsewhere.

If performing your own post-mortem, you should ask your veterinarian to show you how it should be performed and how to identify the major organs. **Always** wear a mask and gloves to prevent exposure to potential pathogens which may infect people, such as *Salmonella* and *Chlamydia*. Perform the post-mortem in a well ventilated area, away from your collection.

- The third option is to have the post-mortem performed by an avian veterinarian. He/she will be familiar with the normal anatomy of your bird and be able to perform a variety of tests to try and identify the causes of a bird's death. Unfortunately, a gross post-mortem examination may not reveal enough information to come to a diagnosis. In these cases, tissue and organ samples from the bird can be sent to a veterinary pathology laboratory for further analysis and diagnosis. Microscopic analysis of tissues (histopathology) may reveal evidence of a virus, toxin, nutritional or metabolic disease.

To benefit most from a post-mortem examination, it is best to wet the bird's feathers down with detergent and then put it in the refrigerator - **not** the freezer. Freezing leads to cellular damage when the body is thawed, making histopathological examination very difficult.

The bird should be taken to the veterinarian as soon as is possible and within 48 hours after death. After this time, decomposition will drastically reduce the effectiveness of post-mortem. If several birds have died over a few days, it may be worth taking more than one to the vet to ensure that the maximum amount of information can be obtained and to confirm that the same cause has resulted in the death of all birds.

The death of a bird is an unpleasant experience. By having a post-mortem performed at least the reason for its death may be found, the appropriate management changes implemented, and if necessary treatment given, hopefully preventing further deaths.

Disease Prevention

The impact of diseases can be minimised if the following steps are taken.

- Purchase healthy birds.
- Quarantine new birds.
- Good aviary/cage design, feeding and general husbandry.
- Regular health assessment.

Purchasing Healthy Birds

The steps on how to go about purchasing a healthy bird have been discussed in previous sections.

Quarantine

As well as observing the birds and looking for signs of disease as discussed previously, there are several tests for diseases which should be carried out during quarantine. Droppings can be examined for internal parasites, especially intestinal worms. This is best done by a veterinarian unless one has been shown how to do this by a professional. If the droppings can't be tested, it may be prudent to strategically worm the birds at the commencement of quarantine, two weeks later and again just before release from quarantine. This will hopefully eliminate any worm burdens present.

Testing for the presence of Psittacine Circovirus Disease (PCD), also known as Psittacine Beak and Feather Disease (PBFD), is very important in cockatoos, particularly with young birds. As will be discussed later, this disease can be equally devastating to the pet bird owner and the breeder, so diagnosing its presence early is imperative. Enlisting the services of an avian vet is required for this, as it is for the detection of Chlamydiosis (Psittacosis), Polyomavirus and other infectious diseases which may be found in birds.

The quarantine period is the ideal time for gradually changing the diet of a cockatoo to that which a keeper desires. This should only be attempted after the bird has had a chance to settle into its new environs. Cockatoos can be notoriously conservative with their diet, although young birds are more receptive to new food items. Perseverance will hopefully result in a broad diet which will minimise long-term nutritional problems to which cockatoos are often susceptible.

The temptation to introduce a newly purchased bird immediately into the aviary without quarantine must be resisted. Remember that quarantining birds is the most effective way of preventing disease from entering a bird collection. It also allows time for new birds to acclimatise to new surroundings and a new owner. This will reduce stress on the birds when finally placed in a new aviary.

Aviary Design and Daily Management

Birds should be housed so that they are protected from climatic extremes and are happy in their environment. Aviary design should allow for easy servicing so that feeding and cleaning can be done as easily and quickly as possible, as this minimises stress.

Preventing build-up of stale food and other organic matter can eliminate sources of microbial and parasitic infection. Constructing aviaries so that wild birds are prevented from landing on top of flights will stop transmission of disease through direct bird-to-bird contact and via the wild birds' droppings.

Other factors, such as solid partitioning between flights to prevent fighting between pairs, will avoid injuries and allow birds to settle down better for breeding.

With pairs kept in aviaries, regular observation will allow early aggression by the cock towards the hen to be detected - a problem encountered in several cockatoo species.

Regular Health Assessment

It is advised to have the health of your cockatoos checked on a regular basis. For aviculturists with pairs of cockatoos, this may be as simple as having faecal tests for internal parasites done on a 6-12 monthly basis. By pooling the droppings from several aviaries together in the one test, this procedure can be done at minimal expense.

Breeding birds should be assessed before the breeding season to ensure that they are not obese. Young birds should be checked for bone strength, proper feather development and powder down production. Of course, more specific tests may be required for specific problems.

Viral Diseases
Psittacine Circovirus Disease (PCD)

This disease, also known as Psittacine Beak and Feather Disease (PBFD), is caused by a virus which infects and kills the rapidly dividing cells of the beak, feathers and the immune system of parrot-like birds.

All parrots are susceptible, however young cockatoos are probably the most commonly seen affected birds. It is widespread in the wild population of cockatoos as well as in captive birds.

This disease most commonly affects young birds, although birds of any age can be affected.

The disease takes on two main forms.

- *Acute Form*. Birds present very ill, often with green or mucoid diarrhoea but without any obvious feather or beak abnormalities. Although the virus can affect the liver, it is the virus's ability to suppress the bird's immune system which makes the birds susceptible to secondary bacterial, viral, chlamydial and fungal infections. This often results in rapid deterioration and death in affected birds.

- *Chronic Form*. The most recognisable form in cockatoos is the chronic form. Birds gradually have their moulted feathers replaced by abnormally formed new ones. These affected feathers are often fragile, are uneven in shaft thickness, may be dried and withered, with blood in the base of the quill. Many fall out prematurely. In cockatoos the powder down feathers are often the first affected. The result is a decrease in the production of the fine white 'feather dust', so the plumage appears darker than normal. Galahs appear darker pink and grey, Major Mitchell's Cockatoos appear darker pink and Sulphur-crested Cockatoos and corellas appear dirty. The normally grey beak and feet of the Sulphur-crested Cockatoo turn black. Bare areas begin to appear, first in the powder feathers, then as the contour feathers are affected, bare areas on the body become more apparent. Eventually, wing and tail feathers fall out. The actual appearance of the bird depends upon the stage of moulting when the bird was first affected.

As well as appearing shiny, the beak may grow excessively long and show cracks on its surface. As the disease progresses, secondary infections of the beak may cause it to rot. Parts of the beak may fall off, making it difficult for affected birds to eat.

Infected birds are also more predisposed to a variety of secondary infections which will eventually lead to their demise.

The virus is shed in the feather dander, faeces and possibly oral fluids. Although not conclusively proven that PCD can be transmitted in the egg, the virus can possibly contaminate the surface of the egg, infecting chicks as they hatch. Incubating eggs in a totally hygenic environment would definitely minimise this risk. Therefore, the greatest risk of infection is in the nest and when young birds are brought together from differing sources, ie during trapping for the pet bird market or in handrearing facilities.

The disease can be diagnosed by laboratory tests which may examine blood, feathers and/or tissue samples from suspect birds. Your avian veterinarian may be able to give you a provisional diagnosis based on clinical examination of the bird alone, before suggesting laboratory tests to confirm the presence of the disease. There is no treatment for PCD and the disease is nearly always fatal.

In birds showing clinical signs of PCD, supportive care in the form of soft foods, vitamins and antibiotics for secondary infections may be helpful in prolonging the life of a treasured pet, but should not be considered if there is a risk of infecting other birds in the collection or household. A vaccine to protect birds from the virus has been experimentally developed but at this stage is not commercially available.

The best way of preventing the disease from entering a collection of birds is to purchase birds from PCD-free flocks where possible. All new birds should

B. RITCHIE

Right: Psittacine Beak and Feather Disease (PBFD) is a slow and cruel virus. The onset of signs is not sudden but affected birds slowly show signs of illness and deterioration. These birds may have deformed or missing feathers and as time progresses the beak will become deformed and very long. The underside of the upper beak will also become rotten. Most commonly affected parrots are Sulphur-crested Cockatoos, corellas and a growing number of other parrot species.

be kept isolated from the resident birds. Quarantining of new purchases and testing for PCD at this time will help prevent introducing the disease into a collection.

The virus appears to be quite stable in the environment. Therefore in an outbreak situation cleaning of the aviary environment, disinfection with appropriate chemicals eg Virkon S™ or glutaraldehyde (Parvocide™), disposing of wooden aviary furnishings (eg perches, nestboxes) and removal of soil/sand etc from the aviary floor may all help to limit the spread of infection. This is a nasty disease of which all aviculturists should be aware.

Avian Polyomavirus

Avian polyomavirus is a viral pathogen which can cause widespread deaths in psittacine nestlings, particularly in handrearing environments. Chicks usually die quickly, but some may show bruising, pale skin, weakness or yellow urates shortly before death. Adult birds usually are resistant to the disease but can become infected and occasionally die.

Polyomavirus has been found in wild cockatoos in Australia. Cockatoos, in general, are highly susceptible to infection but unlike many other parrots rarely develop the disease unless they are suffering from an immunosuppressive disease such as PCD.

Although this virus can cause widespread losses to exposed young parrots, it appears that young cockatoos may only suffer from feather abnormalities and temporary illness and usually recover with supportive care. The main cause for concern in a mixed collection is the possibility of cockatoos spreading the virus within the collection whilst remaining healthy themselves. Testing is now available in Australia and a vaccine does exist in the USA.

In a nursery outbreak situation, there is not much that can be done for infected chicks except to provide supportive care. Hygiene should be improved, birds should be spread out as much as possible and individual feeding syringes should be used for each chick. No new chicks should be introduced to the nursery. Disinfection programs with appropriate disinfectants should be undertaken eg 5% sodium hypochlorite (bleach) at a concentration of 50ml/litre of water, VirkonS™. If possible, an effort should be made to determine the source of the problem.

Raising young from other collections, moving birds to and from outside aviaries, having poor traffic flow back and forth between adult birds and handreared chicks and not quarantining new arrivals all increase the risk of avian polyomavirus entering the collection.

Chlamydiosis / Psittacosis

This is a disease caused by the organism *Chlamydia psittaci*, which can infect cockatoos and other parrots as well as other birds and mammals. It can be very contagious and can cause a wide range of clinical signs from mild illness through to severe disease and sudden death. Some of the more commonly seen symptoms include mild depression, lethargy and rough plumage, sneezing, inflamed watery eyes, yellow-green or watery droppings and diarrhoea, weight loss and occasionally tremors and fits.

The organism can cause disease in its own right, or be secondary to other diseases which have already lowered the bird's immune system, such as PCD, or stress due to transportation and new accommodation, moulting, weather extremes etc.

The organism is spread in the bird's droppings, eye or nasal discharges and feathers and can last up to eight months in aviaries where faeces are allowed to build up over long periods. One problem with this disease is that apparently normal, healthy looking birds can carry *Chlamydia* in their system and only shed it intermittently. These 'carriers' may then break down and become ill with the disease or just act as a source of continual reinfection for other birds.

Diagnosing the disease is usually achieved by a combination of physical examination and laboratory testing. However, there is no diagnostic test available which can guarantee that a bird is free of this disease.

Treatment of this disease is usually done by long-term treatment with tetracycline antibiotics. Doxycycline (Psittavet™) is the drug of choice. However, at the end of this treatment, birds are still susceptible to reinfection and some birds become life-long carriers. It may therefore, be wise to cull particular birds which consistently resuccumb to this disease.

Stopping the spread of chlamydiosis in a collection involves isolating and treating infected birds, cleaning of aviary floors regularly to remove sources of reinfection and using a suitable disinfectant to clean aviaries and aviary utensils during treatment, eg in-contact birds may, at the veterinarian's discretion, also be treated with antibiotics prophylactically. Any stresses such as bullying, breeding or poor diet must be corrected at this time. New birds should not be purchased during this period.

Be warned! Chlamydiosis is a zoonotic disease, ie it can be contracted by humans.

In order to minimise the risk of contracting the disease, cages should be cleaned in well ventilated areas and safety precautions such as wearing masks and gloves when attending to ill birds should be practised.

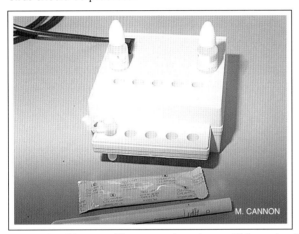 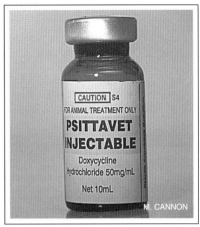

Left: The Clearview™ test used to detect Chlamydia (Psittacosis) in sick birds. It is not as useful in birds that are not sick.
Right: Psittavet™ - the injectable form. One of the best antibiotics for Psittacosis.

Bacterial Infections

Bacteria can cause a wide variety of diseases in all birds, including cockatoos. Many of the signs shown by birds with bacterial infections can mimic those of other diseases.

Birds can be lethargic, not eating and fluffed, have eyes closed, discharge from nostrils and eyes, sneezing, laboured and/or noisy respiration, tail bobbing, diarrhoea (often bright green in colour), soiled vent, excess urine production, weight loss and even sudden death.

Bacterial infections can be localised to particular body systems to cause specific diseases, (eg upper respiratory tract disease causing blocked nostrils, swollen sinuses and sneezing or intestinal disease causing diarrhoea), or they can cause generalised disease throughout the body.

There are many different species of bacteria and often many strains within the one species. These bacteria and their strains can all vary as to what type of disease they cause and to what antibiotics they are sensitive. Therefore, there can be no one antibiotic that is effective against all bacterial infections.

Using the correct antibiotic is only one part of the treatment plan needed to deal with bacterial infections. When confronted with a bacterial problem in an aviary, the first reaction of most bird keepers is to rush and put the 'best' antibiotic they can find in the water source. What we need to realise is that in our search for a quick-fix answer we are overlooking the real root of the problem, that being a breakdown in our avian husbandry.

Bird keepers should concentrate on improving that which is under their control - namely **good management**. Good management means keeping birds on a healthy, nutritious diet and in a clean, stress free environment. The following are just some of the questions we need to answer to help pinpoint potential sources of bacterial problems within our aviaries.

• Are the cage floors filled with faeces, decaying food etc? These provide an ideal environment for many bacteria to survive and replicate.

• Is food and water clean, of good quality and provided fresh daily? For example, rainwater collected in tanks and PVC watering systems have been found to contain *Pseudomonas* sp. bacteria.

- Are nestboxes cleaned out or replaced regularly?
- Are mice present in the aviary or allowed to contaminate food? Mice can carry such nasties as *Yersinia* sp. and *Salmonella* sp. which can cause widespread deaths in aviaries, particularly during the cooler months of the year when mice are more likely to move into aviaries.
- Do wild birds regularly perch on top of the aviary?
- Have new birds been recently introduced into the aviary? Lack of adequate quarantining of new birds is often responsible for introducing disease.
- Are birds stressed by being kept in draughty aviaries, with incompatible mates which intimidate them, near areas of high traffic flow to which they are not accustomed, where the population density is too high eg too many young birds crowded in the one aviary after a successful breeding season?

As can be seen from these questions, prevention of disease by maintaining good management practices is the ideal way of minimising the impact of any disease agent. Maximising a bird's immune system and decreasing its exposure to the disease agent is the ideal way of maintaining the health of the birds in our care.

Bumblefoot

Bumblefoot is an inflammatory or degenerative condition of the avian foot, commonly seen in pet cockatoos.

In early cases the skin on the underside of the foot appears thin and smooth instead of having a knobbly texture. Flaky skin may also be present. As the disease progresses the underside of the foot becomes redder and eventually infected. The infection firstly involves the soft tissue of the foot but in severe cases can spread to the bones. Clinical signs of the disease can vary from very slight changes in perching posture to more obvious leaning to one side when perched, to total disuse of the affected foot.

There are believed to be several causes of this condition. In many cases malnutrition may predispose a bird to this condition. Cockatoos fed a predominantly seed-based diet may be Vitamin A deficient which leads to a weakening of the skin structure. Many of these birds are obese and don't exercise adequately which means that a lot of weight and pressure is placed on the skin under the feet. This exacerbates the existing weakness in the skin structure, resulting in erosion and ulceration of the foot pads.

If the perching is too smooth or too rough, or is of too small or too large a diameter, then this only hastens the condition as there is an unevenness of weight distribution across the foot.

Sandpaper perches should be avoided as they may increase abrasions on the already weakened skin of the foot. If the problem occurs in one foot and the bird spends increasing amounts of time on the other foot, then this second foot is put under increased pressure and may also succumb to bumblefoot.

Treatment will vary with the severity of the lesions. In very early cases, supplementation with multivitamins (especially Vitamin A and Biotin) in the short-term and changing the bird's diet long-term, to reduce obesity and improve vitamin, mineral and amino acid levels and balance, should be made.

Perches should be altered to provide a more variable diameter of natural perching material, eg branches from non-toxic trees, such as eucalypts, melaleucas etc. Avoid smooth dowel, metal or plastic perches and discourage birds from hanging on the cage wire for long periods. Increasing the level of the bird's exercise is also of benefit.

In cases with small wounds, as well as the changes suggested, birds are usually treated with antibiotics and the perches may be padded. Topical treatments may also be of assistance eg Betadine™ solution.

Fungal Infections

The two most commonly encountered fungal infections in cockatoos are aspergillosis and candidiasis.

Aspergillosis

Aspergillosis is primarily a disease of the respiratory system in birds, although other body systems can be involved. The disease can cause two main disease entities. The first is where the birds show obvious breathing difficulties with open-mouthed breathing and tail bobbing, particularly after exercise.

They may be depressed and have obvious weight loss suggesting that the disease has been occurring for some period of time before the breathing difficulties became noticeable. Many of these birds may also have green stained droppings.

The second form is more sudden with the bird showing a sudden onset of obvious breathing difficulties without weight loss. Many of these birds may die quickly without treatment. In fact, some are just found dead without any previous signs of illness.

The *Aspergillus* organism is found commonly in the environment and is breathed in by most animals and people everyday. Disease is thought to occur either through exposure to very high numbers of fungal spores or due to immunosuppression (with the latter being the most common). Stresses such as transport and handling, poor ventilation, inadequate nutrition, very young or old age, abuse of antibacterial or corticosteroid medications, respiratory irritants and of course, underlying disease, such as PCD, may predispose a bird to developing aspergillosis.

High numbers of fungal spores can be found wherever there is rotting organic matter. Spilt feed, seed husks and stale green food, particularly when exposed to moisture and warmth, can all support high levels of fungal growth. Perhaps the most risky situation is where a bird is exposed to a combination of these factors, ie stress and high spore levels. Examples of this are hens working dirty, contaminated nest hollows or birds transported in carry cages with spoilt hay or straw bedding.

The history and physical examination may suggest an *Aspergillus* infection in more obvious cases, but obtaining an exact diagnosis in the live bird can be difficult. Treatment of this disease definitely requires veterinary treatment.

Preventing this disease from occurring is far preferable to trying to treat it. Optimising nutrition, hygiene and housing, and eliminating sources of infection are very important. In particular, paying attention to cleaning aviary floors, nest logs and carry cages regularly and eliminating wet spots in the aviary are all important.

Candidiasis

Candidiasis is caused by the yeast *Candida albicans*. This organism is common in the environment and may be a normal inhabitant of a bird's digestive tract. Its ability to cause disease depends upon the age of the bird and its immune status and the presence of other causative factors.

In young birds whose immune systems are still not fully developed, candidiasis is frequently seen, especially in handreared birds. Usually, some fault in the handrearing environment has allowed the organism to cause a problem. Examples of this are fluctuating brooder temperature or humidity, problems with decomposition, temperature, texture or nutritional content of feeding formulas, food caked onto the chick's face and plumage and abuse of antibiotics in chicks. It can cause disease in its own right or become a problem as a result of another disease. In very severe cases, the disease can spread right throughout the digestive tract and is very difficult to treat.

Affected chicks usually regurgitate or vomit, have slow crop emptying times, are depressed and refuse food. In some cases the crop may become impacted. In older birds the crop may be full of mucous, and crop emptying is often delayed. A sour smell is often associated with this. Many of these birds may have underlying Vitamin A deficiencies which may predispose them to infection with *Candida*.

In all infections, birds may show raised white areas lining the inside of the mouth, crop or other affected area of the gastrointestinal tract. Occasionally, yeast infections can also occur on the skin, feet and respiratory tracts.

Diagnosing candidiasis involves identifying the organism by staining a smear of the affected area and examining it under the microscope.

Treatment of this disease firstly involves correcting the underlying management or other disease problems and then treating with appropriate medications. Many aviculturists are familiar with nystatin (eg Nilstat™, Mycostatin™) for treating this disease. With severe or non-responsive cases, antifungals such as ketoconazole (Nizoral™) may need to be used but these are more likely to have side effects, especially in chicks. Birds with candidiasis may also need supportive therapy, eg fluids, Vitamin A.

It is important for handrearers to realise that if they are having continual problems with candidiasis, they must critically reassess their handrearing techniques and daily chick management.

Parasitic Worms

There are many species of parasitic worms which can infect birds. Only the most common species found to infect cockatoos are discussed here.

Above: Candida, as can be seen in the yellowish area in the photograph, can be caused by unhygienic handraising methods.

Roundworms (Ascarids)

These are the most common worms seen in parrots. Adults are long, creamy white and smooth and are usually seen in a bird's droppings after worming or in the intestine during post-mortem examination. Worms are rarely seen in a bird's droppings, unless the infestation is severe.

Birds with worm infestations may show weight loss, lethargy, have diarrhoea, be more susceptible to other diseases and have poor breeding results. In severe cases they may be emaciated and even pass seed in their droppings. Some birds may die suddenly. Others may not appear ill at all but may shed many eggs in their droppings.

This worm is transmitted by direct contact with infected droppings either from the same or another infected bird. To prevent reinfection, droppings have to be removed regularly or birds should be housed so that they can't access their droppings, eg suspended cages. The eggs of roundworms are very resistant in the environment where they can persist for years especially in moist organic matter. Disinfectants are ineffective in destroying these eggs. Therefore preventing infection from occurring in the first place by quarantining and treating newly acquired birds, having dry floors which can be easily cleaned (eg sloping concrete) and fully roofed aviaries so that wild birds can't infect an aviary should all be considered.

Diagnosing a worm problem in a live bird is best done by examination of the faeces for worm eggs. These are not visible to the naked eye and need to be checked microscopically. It is always wise to interpret the results of a faecal examination with the bird's condition in mind and in consultation with an avian veterinarian.

Treatment of worms is best done by administering one of the wormers listed in the worming table on page 62 either direct to the mouth or by crop needle. In-water medication is very unreliable as many factors can vary the volume of water that a bird will drink, eg taste, weather conditions, breeding status. Therefore, underdosing commonly results. In hot weather overdosing can be a problem.

The frequency of worming will depend upon each aviculturist's situation. Where infection is already present in an aviary, where there is a dirt floor and open roof and the birds in question spend much time on the ground, then worming may have to be carried out regularly, in extreme situations as often as every six weeks. Perhaps the most important times are before breeding and then when youngsters fledge as this is the time they spend longer periods on the floor and are under stress. Then again, if worms are not a problem within aviary there is no need to treat for them. Indiscriminate use of worming treatments (anthelmintics) is at the least wasteful and at worst promotes resistance to these drugs. Having droppings tested is not expensive and can save unnecessary medication.

Threadworms or Hairworms (Capillaria)

These are tiny thread-like worms which can be difficult to see with the naked eye. They can cause diarrhoea, weight loss, lack of appetite, vomiting and anaemia. They can live from the oesophagus and crop through to the intestinal tract.

The life cycle begins when a bird swallows a hairworm egg which is infective (ie has been on the ground for approximately four weeks and contains a larva). Once swallowed, the larva hatches from the egg and burrows into the gut lining and develops into an adult worm. The eggs can remain infectious in the environment for several months. Earthworms may carry a larval stage of the hairworm so should be eliminated from the environment.

This worm can be difficult to diagnose as eggs may only be shed intermittently. It can be resistant to the more traditionally used worming medications. In these cases, moxidectin (eg. Cydectin™ Sheep Drench) may be effective.

Tapeworms

These are long, flat and whitish worms whose bodies are usually made up of many segments. They are most commonly seen in wild caught cockatoos and only occasionally in aviary bred stock. In large numbers they can cause ill thrift and diarrhoea, although many birds may appear to be normal.

The life cycle begins when the end segment of the adult worm which contains eggs is shed in the droppings. An egg is then eaten by an intermediate host such as an insect, earthworm, snail etc. The intermediate host is then eaten by the bird and then the tapeworm matures and attaches directly onto the intestine. The segments are infrequently shed in the droppings but sometimes segments can be seen hanging from the bird's cloaca after defecating. Treatment is best achieved with praziquantel (eg Droncit™, Avitrol Plus™) and by eliminating access to insects.

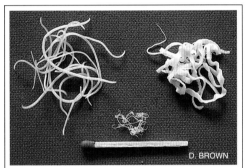

Above: Worms. Top right - Tapeworm;
top left - Roundworm;
bottom - Gizzardworm.

Worm Treatments Used in Cockatoos

The following is a list of some of the worm treatments used to treat worm infections in cockatoos. Please remember that of those listed, only Avitrol™ and Avitrol Plus™ are registered for use in birds and the manufacturer's instructions have been given here. The other parasiticides are registered for other species of domestic animals. As such, all responsibility for the use of these drugs is totally in the hands of the user. In general, it is inadvisable to worm birds in extremely hot weather, or to worm birds that are stressed or feeding young.

- **Avitrol™, Avitrol Plus™**. These products contain levamisole which is effective against roundworm and some strains of hairworm. Avitrol Plus™ also contains praziquantel and will therefore also treat tapeworm. The directions for both products are to give 1ml direct to the crop per 240gram body weight. This dose should be repeated in 10-14 days. It can be given orally but is bitter and the birds may regurgitate. Accurate dosing is important as overdosing can lead to toxicity problems such as lack of coordination, leg and wing paralysis and even death. In-water dose rates are given as 12 drops to 20ml or 25ml per litre of drinking water.
- **Panacur 25™**.This product has been used for many years by aviculturists and is recommended for treating roundworms and some strains of hairworm. The suggested dose rate is 1ml to 2mls per 500gram body weight, ideally for three consecutive days. Dosing should be repeated in 10-14 days. Being a suspension, it should be shaken well before use.

It settles out in water so is not entirely suitable for in-water use. A dose of 5ml per litre of soaked seed fed for five days has also been recommended. It should not be used in birds when moulting as it may cause feather deformities. Another product Panacur 10™ has been associated with deaths and should not be used even when diluted.

- **Ivermectin**. This is available as several preparations. It is suitable for treating roundworms and some capillaria. The effective dose rate varies from 0.2mg per kg body weight up to 0.8mg per kg body weight in resistant cases. Dosing should be repeated in 14 days. The dose given in the worming table is for Ivomec™ Liquid for sheep (0.8 gram per litre) which has been diluted 1 in 10 with water immediately before use.
- **Moxidectin**. This is related to the ivermectins but is particularly useful for resistant hairworm infections. It is available as various forms of Cydectin™ drenches for sheep and cattle. The correct dose for birds is still being investigated but a dose of 0.2 mg per kg body weight has been suggested. It also causes vomiting if placed directly into the beak. Using a clean crop needle direct to the crop is the preferred method of administration.
- **Pyrantal pamoate**. This is quite a safe product used for treating worms under such trade names as Canex Puppy Suspension™. A dose of 4.5mg per kg body weight orally repeated in 14 days is recommended. Its main effect is in treating roundworms.
- **Praziquantel**. Most commonly available as Droncit™ tapewormer for dogs or cats, Virbac™ tapewormer and is also found in Avitrol Plus™. It is the drug of choice for treating tapeworm infestations. The dose rate is 10mg to 20mg per kg body weight repeated in 10-14 days.
- **Other wormers**. Other products such as oxfendazole (Synanthic™, Systamex™) and netobimin (Hapavet™) have also been used successfully by some aviculturists and veterinarians. The doses for Synanthic™ and Systamex™ are 2ml per litre of water for three days, repeated in 14 days. For Hapavet™, the directions on the container should be followed.

DIRECT DOSES OF SOME COMMON WORMING MEDICATIONS FOR COCKATOOS

COCKATOO	AVERAGE WEIGHT	PANACUR 25™	AVITROL PLUS™	IVOMEC™#
Galah	340grams	0.7 - 1.3ml	1.4ml	1.0ml
Short-billed Corella	550grams	0.9 - 1.8ml	1.8ml	1.2ml
Eastern Long-billed Corella	640grams	1.3 - 2.6ml	2.7ml	1.7ml
Western Long-billed Corella	650grams	1.4 - 2.8ml	2.9ml	1.8ml
Major Mitchell's	420grams	0.8 - 1.6ml	1.6ml	1.0ml
Sulphur-crested	870grams	1.8 - 3.5ml	3.6ml	2.2ml

Weights of individual birds may vary widely. Where possible birds should be weighed before treatment, especially with levamisole based wormers.

\# These doses are for Ivomec Liquid™ for sheep (0.8g/L), diluted 1 in 10 with water immediately before use.

Disclaimer.
Very few drugs are registered for use in birds, and most usages and dose rates have been extrapolated from mammalian therapeutics. Everyone using medications should be aware that manufacturers of these drugs will not accept any responsibility for the 'off-label' use of their drugs. The above dose rates and information are based on clinical trials and practical experience, but unrecorded adverse side-effects may occur. Where possible, the author has provided brand names for the drugs mentioned. These should not be taken as a recommendation for one particular brand over another; but rather as a starting point for you to find the drug of your choice. In most instances, contra indications (C/I) and side-effects are not listed. This should not be taken to mean that there are none - many of these drugs have not been used extensively, and reports on contra indications and side-effects are not recorded at date of publication.

Protozoan Parasites

Protozoans are single-celled microscopic organisms. There are several species which are parasites of the digestive tract of birds. The one most likely to cause problems in Australia is *Giardia* spp.

This parasite may cause a smelly mucoid diarrhoea, anorexia, depression, weakness and death, particularly in young or immunosuppressed birds. In cockatiels in the USA, *Giardia* has been associated with feather plucking and self mutilation. Some birds may carry the organism but appear normal.

Microscopic examination of **fresh** faeces or scrapings from a dead bird's intestine are required for a diagnosis.

Treatment involves administering the appropriate antiprotozoal drugs, eg metronidazole (Flagyl^{TM}) or ronidazole (Ronivet-S^{TM}) for approximately seven days and supportive care with fluids, antibiotics and/or antifungals, multivitamins etc.

The environment must be kept dry to prevent survival of the cystic stage of the organism's life cycle and hence prevent reinfection.

There is some conjecture as to whether the parasites causing infections in birds can also cause illness in people. Care should be taken when handling infected birds and when cleaning their cage or aviary.

External Parasites

There are many species of lice, mites, ticks and flies which can infect birds including cockatoos.

Scaly mites (*Cnemidokoptes* spp), which are well-known for causing scaly face in Budgerigars and tassle foot in Canaries, may occasionally be seen in cockatoos. The mite burrows into the skin producing a crusty reaction which is visible as a honeycombing of the skin especially around the eyes, beak and face but potentially anywhere around the body. The mite is spread by direct contact or through skin flakes shed from an infected bird. Killing the mite is relatively simple with ivermectin or moxidectin being effective. However, in nearly all cases seen to date in cockatoos, there has been an underlying immunosuppressive disease in these birds. Most of these birds have in fact been infected with PCD. Therefore, controlling the mite in these circumstances is more complicated.

A variety of other mites and lice can be found on birds. The eggs of lice can often be seen on the underside of feathers as a fine, white or grey powder arranged along the shaft of the feather.

The adult lice are usually seen scurrying away from the light. When in large numbers, they may climb onto the handler and cause short-term irritation. Large numbers of lice on a bird also suggest that its immune function is poor and the bird should be checked for PCD.

Mites are much smaller and may appear as a brown-red fine powder on the underside of the bird's feathers or on its body. Some mites actually live within the shaft of the feathers.

Birds with mite and lice infestations may show skin irritation, excessive preening, poor feather quality, abnormal moulting and in severe cases anaemia. Again, bird specific insecticidal sprays and powders may be used successfully to treat these birds.

Cleaning the cage environment and treating with an insecticide (eg Avian Insect Liquidator™) may also be necessary in cases of mite infestations.

Pigeon fly are flattened flies which live on the cockatoo's body and often crawl out onto the owner when the bird is handled or even fly into the owner's hair! They are generally considered harmless but may be involved in the transmission of blood parasites.

Heavy Metal Toxicity

This is a common problem in pet cockatoos caused by the ingestion of substances containing heavy metals, in particular zinc and lead and to a lesser extent, copper and mercury.

The most common source in aviary birds is galvanised coating on bird wire which contains 98% to 99% zinc and 1% to 2% lead. Wherever the galvanising forms chunks on the wire (eg near weld points) or whenever the birds chew the wire, heavy metal toxicity is a potential problem. The shinier the wire, the higher the zinc content.

In pet birds, the sources are more varied and can include old paint, jewellery, leadlight windows, curtain weights, bird toy weights and galvanised feed or water dishes.

Affected birds will often drink excessively and pass watery droppings, have greenish diarrhoea, show vomiting or regurgitation and slow crop emptying. Others may just appear depressed and lose weight. In some instances the bird may seizure. In chronic, low-grade exposure, fertility may be decreased. Some birds may also feather pick or chew their feet.

Initially, affected birds will usually require hospitalisation to stabilise their condition. X-Rays and blood tests may be undertaken to confirm a diagnosis. Delaying of treatment can be disastrous - the aim is to minimise the amount of heavy metal toxin the bird absorbs. It is difficult to reverse the damage already done by the toxin. If a bird is quickly treated and stabilised, it may then be possible to send it home with treatment. If, however, the bird is severely affected and blood can be seen in the droppings, the chances of a successful outcome are very low.

Prevention is the best way to deal with this disease. This involves being aware of potential sources of heavy metal in the bird's environment and eliminating or at least minimising its exposure to these sources where possible. Scrubbing the wire with a vinegar solution and wire brush thoroughly will help to dislodge loose fragments and remove 'white rust' but will not prevent it from reforming.

Contrary to popular belief, 'weathering' the wire by leaving it exposed to the sun and rain for six months or longer does **not** render the wire safe. Where possible use top quality wire as this is cleaner. Reject wire with many lumps forming at the weld points. BHP's new Evencoat™ galvanised wire has virtually eliminated the lumps at weld points that have been a problem in the past.

With pet birds, these should always be supervised when left free in the home and all potential heavy metal sources should be removed or covered.

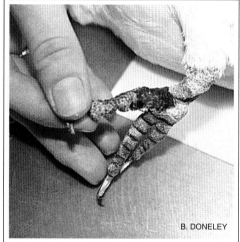

B. DONELEY

Above: Cockatoo with a mutilated foot.

Aggression and Traumatic Injuries

Many cockatoo cocks have a reputation for attacking their mates causing serious injury or even death.

These attacks are often predictable. The popular theory is that the cock comes into breeding condition before the hen. As he tries to court her and she ignores him, he becomes frustrated and attacks her and within the confines of an aviary can inflict serious injuries. However, these attacks can occur at any time of the year, not just during the breeding season. It is not only newly formed or incompatible pairs which may exhibit this behaviour. Unfortunately, the cocks of some long established pairs which have successfully and peacefully reared young over a long period of time may suddenly turn on a hen and injure her.

Most injuries occur to the hen's beak, eyes and face. Occasionally other areas of the body such as the feet may be involved.

There are several things that can be done to help minimise this from occurring.

- Closely observe pairs to detect early signs of breeding activity or aggression towards the hen. Closed circuit TV monitors can assist in observing nervous or secretive birds.
- Decrease the cock's mobility, eg clip one or both wings before the start of the breeding season. Only clip enough to slow the cock bird's advances. The problem with this procedure is that it may not be easy to predict when the cock is coming into breeding condition. In addition, wing clipping does not stop a bird from climbing up into the breeding box or log and cornering the hen in the nest.
- Remove the cock from the aviary. He may be housed next to or near the hen and when the hen is seen working the log or calling the cock, he may be reintroduced and closely monitored! Under **no** circumstances should the hen be removed as her reintroduction to the cock's aviary

will almost certainly result in disaster. If the hen has to be removed eg for treatment, then the two birds should be introduced simultaneously into a new aviary.

- The cock should be re-paired with a new hen in a new aviary.
- Try to allow potential breeders to choose their own mates at an early age. This maximises the chances of achieving compatible pairings and therefore breeding success.
- Remove the nestbox from the aviary.
- Use nestboxes with more than one escape door. This is particularly helpful for hens that hide in nestboxes.
- Some cocks may be aroused by neighbouring pairs, particularly where only wire separates them. Relocating pairs to a new part of the aviary complex may allow agitated cocks to settle down.
- Maximising cage sizes allows the hen more chance to escape. In some instances double aviaries adjacent to each other with one small access point at either end connecting the two may be useful.
- In the USA, some veterinarians have experimented with acrylic prostheses attached to the end of the cock's beak. This does not allow him to bite the hen. The prostheses last for several weeks to months but eventually fall off.
- Filling the nestbox with blocks of soft, untreated timber can divert the cock's aggression from the hen to preparing a suitable nesting site.

Dealing with Injuries

If a wounded bird is discovered it should be immediately removed from the aviary to prevent further injury. The bird must be quickly assessed to determine whether it can be treated at home or whether it requires immediate veterinary attention. Bleeding needs to be stopped and the bird allowed to recover from its state of shock. Most birds will be stressed, therefore transferring them to a warm, quiet and slightly darkened environment is the best first course of action. Some will have lost considerable blood and will require warm fluid therapy (eg Hartmann's Solution™). This is best given by injection into the vein or under the skin as most attacked birds will have damage to their beaks and mouths. Wounds need to be cleaned and then treated with a topical antiseptic such as Betadine™ or chlorhexidine (eg. Hibiclens™). Antibiotics may also need to be given. Your avian veterinarian can best advise you of the best course of action for any particular case.

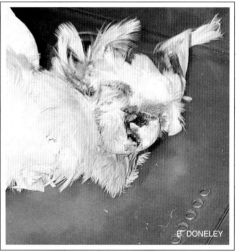

Above: Beak trauma in Major Mitchell's Cockatoo hen caused by her mate.

Nutritional Problems and Diseases

One of the commonest reasons for illness in cockatoos is improper nutrition. This has resulted from the widespread practice of feeding these birds on an almost exclusively seed-based diet, in particular one with a high percentage of sunflower seeds. All-seed diets are not nutritionally balanced no matter what the combination of seeds. They tend to have an excess of fat, be low in protein levels with poor amino acid balance and are often deficient in minerals, trace elements and vitamins.

It is important to remember that cockatoos have largely adapted to an arid or semi-arid environment in which they forage for long periods on foodstuffs which are not nutrient rich.

In captivity, we provide them with an ad lib supply of energy-dense food. Their level of exercise is much less than that of their wild counterparts. To make matters worse, many birds will select only one or two seed types from within a mix and feed almost exclusively on these.

Again, sunflower seed is at the top of the list. Its high fat content (47%) makes it very palatable. The result of this type of diet is long-term malnutrition.

Obesity is very commonly seen, particularly in mature Galahs and Sulphur-crested Cockatoos. The most common reason for them being taken to a veterinary clinic is the presence of large deposits of fat under the skin, particularly in the lower abdomen and around the feet. These growths (lipomas) may actually become traumatised due to their rubbing on the perch or cage floor.

It is often the bleeding that brings the problem to the owner's attention. Birds may also bite at these lesions as they may irritate the bird. In these cases, the lipomas may need to be surgically removed.

Long-term treatment involves dietary correction to reduce the fat level and improve vitamin, amino acid and mineral balance. Birds may be converted to a good quality pelleted diet or, at the very least, one with limited or no oil seeds and added vegetables, pulses, fruit and green food. This dietary change needs to be done gradually and with commonsense so as not to starve the bird.

Some birds are very set in their feeding habits and owners may need to be ingenious and persistent to convert them to the new diet. Other birds convert easily with a minimum of fuss. Refer to *Converting Birds to a New Diet*, page 35.

Many of these obese birds also have fatty livers, in which excess fat accumulates in the liver. This can predispose birds to liver failure and spontaneous rupturing and bleeding of the liver, resulting in death. It can also predispose the bird to other problems due to poor liver function. If the liver is greatly enlarged, it may restrict the function of the respiratory tract causing breathing problems.

Other potential problems associated with excess fat in the diet include diarrhoea, infertility, oily feathers and interference with the absorption of some nutrients, such as calcium. As with humans, obese birds are more susceptible to atherosclerosis and heart disease.

There is no truth in the rumour that sunflower seed contains an addictive substance. Its high palatability is believed to be due largely to its high oil content.

Vitamin A deficiency is another commonly encountered problem in pet cockatoos. Seed diets provide little or no Vitamin A or its precursor beta-carotene. Vitamin A is needed for the normal formation of skin and of the tissues which line the gastrointestinal, respiratory and urogenital tracts. It is also needed for normal growth, vision and for red and yellow pigment formation in feathers.

Vitamin A deficiency may therefore be responsible for many illnesses in birds including poor immune function and poor breeding results including early deaths of embryos. Some of the more common syndromes seen in pet birds with this problem include bumblefoot and upper respiratory tract disease (eg sneezing, swollen eyes and sinuses, blocked nostrils). Vitamin A deficiency in these cases may weaken the normal structure and immunity of the tissues in question and allow infectious agents to enter and cause disease more easily.

Good dietary sources of Vitamin A include green food, red peppers, carrots and sweet potato. Cod liver oil is very rich in Vitamin A (and Vitamins D and E) but should be used with caution due to potential rancidity problems caused by improper storage which can deplete the Vitamin E present. It also contains high levels of gizzerosine which can cause gastric ulcers. On the otherhand, over supplementation with Vitamin A can be toxic, particularly since this vitamin can be stored in the liver.

Perhaps the most common nutritional deficiencies seen in cockatoo chicks and breeding birds are due to imbalances in calcium, phosphorus and Vitamin D3 or due to an improper calcium: phosphorus ratio in the diet.

Rickets is a disease which results in improper bone formation in growing chicks so that the beak and skeleton become soft and rubbery.

All seed diets are notorious for having high levels of phosphorus but low calcium levels. Oils in seeds can bind the calcium in the diet making it unavailable to the body, thus worsening the problem.

If bone deformities are detected early when the skeleton is still soft, they can be corrected either surgically or by corrective splinting and the chick has a good chance of growing into a normal adult, once its diet is also corrected. In advanced cases, however, the bones may have broken and rehealed incorrectly and by the time the problem has been noticed, the chick is quite advanced, the skeleton has hardened and corrective surgery may only be partially helpful.

The second problem is known as **secondary nutritional hyperparathyroidism** (SNH).

This is where the body uses calcium at a faster rate than what it can absorb from the intestine leaving the bird weak, thirsty, lacking appetite and sometimes regurgitating. In growing chicks it can lead to bone fractures and deformities. In breeding hens it can cause egg binding, soft-shelled eggs and fragile bones. High phosphorus levels or low Vitamin D levels in the diet inhibit the absorption of calcium from the intestine, thus worsening the condition. These conditions can be treated and prevented by correcting the dietary imbalances.

The normal dietary calcium: phosphorus ratio required by birds is 2:1, with an available calcium level of 0.8%. By contrast the ratios in most seeds varies from 1:37 in corn, 1:7 in sunflower and 1:6 in millet.

Thus, these diets need to be supplemented with calcium and/or Vitamin D3 (Vitamin D3 is needed for proper calcium absorption by the body). The quickest response is had with injectable calcium, but it can cause problems with heart rate if not given carefully. Oral liquid calcium eg Calcivet™ or Sandocal™, is usually better absorbed than powdered forms. Liquid calcium can be given on seed, in water or directly to the bird. Calcium powder can be sprinkled on seed, especially soaked seed or soft foods to which it sticks better.

Many cockatoos will destroy cuttlefish bone without actually ingesting much of the mineral in question. High calcium supplementation should only be carried out during the period of increased calcium demand (eg around egg laying) or during growth or recovery from a deficiency. Over-supplementation can also cause problems.

Vitamin D3 is obtained by exposure to direct sunlight, or in pelleted foods or good quality vitamins made for birds. Fish oils and egg yolks are other good sources.

Birds kept permanently indoors may benefit from using 'full-spectrum' lighting which produces ultraviolet light needed for Vitamin D3 synthesis. These need to be kept no more than 60cm from the bird in order to be effective.

Be aware that excess levels of this vitamin can cause drastic changes in birds including mineralisation of kidneys and other body organs and death. This has been particularly seen in oversupplemented young handreared birds (ie fed Vitamin D3 and calcium supplements on top of an already balanced handrearing formula).

The take home message is that by feeding a balanced diet, many of these problems can be prevented from occurring. Supplementation should be carried out only when required, in carefully considered amounts and only for as long as is needed. It is best to seek the advice of an avian veterinarian or avian nutritionist when considering dietary supplementation.

Feather Plucking and Self Mutilation

Feather plucking is really a general term used to describe any damage done by a bird to its own feathers. If this damage extends to the bird's skin, muscle or other tissues then the term 'self mutilation' is generally used. In actual fact, feather plucking is a form of self mutilation. It is important to realise that these terms describe a whole series of problems, all with different causes and different clinical signs, the common presenting factor being that the bird is inflicting damage to its own body or feathers.

As such, many feather pluckers may look the same, but the underlying causes may be very different. This presents a challenging situation for both the owner and the avian veterinarian. Feather pluckers may just nip the tips off feathers, strip the vane of the feather, split the feather right down the shaft, break the feather off anywhere along its shaft or pull the entire feather from its follicle. This may happen all over the body or be concentrated on wing and tail feathers or just over the chest.

If feathers on the head are affected, then the problem is not one of self-plucking. Either another bird is doing the plucking or the bird has some underlying disease affecting feather health. Careful examination of the feathers should reveal whether a feather has been bitten or has just died.

It is important to distinguish feather disease (eg PCD) from mate-plucking and self-plucking.

Self mutilators also usually target specific areas of the body. Sulphur-crested Cockatoos often chew the muscle of the upper chest and can leave terrible wounds.

There are many potential causes of feather plucking and self-mutilation. Infectious causes include bacterial and fungal infections of the skin, external parasites such as mites and lice and possibly intestinal worm infections. In the USA, feather picking has been seen in cockatiels suffering from intestinal giardiasis. In Australia, when an owner notices their bird plucking, they are often advised by pet shops to treat them for lice infections. Although lice can cause feather problems, they are in truth only responsible for a small proportion of the feather plucking cases seen.

Birds may pluck due to the irritation caused by poor skin and feather quality due to nutritional problems. For example, they may have dry, flaky skin due to deficiencies in Vitamin A, amino acids, biotin, folic acid or salt. In these cases, a change in feather colour may also be seen. Brittle feathers may be due to mineral deficiencies. White streaks in feathers may be due to Vitamin B deficiencies.

In Galahs, malnutrition may show as a change in feather colour from grey to pink. Any disease that affects the body's ability to metabolise or extract these nutrients, eg intestinal or liver disease may also cause poor skin and feather health and hence predispose the bird to feather plucking.

Behavioural causes of feather plucking can often be the most difficult to identify. These include
- Boredom, eg from cage confinement and unlimited food availability.
- Frustration, eg due to unsatisfied sexual needs during the breeding season.
- Anxiety - due to the presence of perceived threats, eg dog, cat or person the bird dislikes, separation from person to whom the bird has bonded or incompatible mate.
- Insecurity, eg keeping bird in an all wire cage, in high traffic areas with no area to hide, keeping birds below shoulder level.
- Changes in a bird's routine or environment.
- Excessive or insufficient hours of daylight.
- Poor socialisation.
- Aggression and teasing.

Feather plucking has also been associated with heavy metal poisoning, Chlamydiosis (Psittacosis) and fatty liver disease.

Sources of pain such as when injuries or fractures occur may predispose the bird to feather plucking and self mutilation at the site of the pain. Feather cysts, tumours, abscesses and localised infections may induce self mutilation. Wing clipping, if done incorrectly can also start the bird on the feather plucking cycle.

Recent evidence suggests that allergies may also be responsible for feather plucking in cockatoos. These allergies can be due to certain seeds or their dust, dust mites, grasses etc. This is an area currently undergoing much research.

The treatment for feather plucking birds will depend on identifying the underlying cause(s). This can be quite lengthy and unfortunately there are no short cuts to be taken if the real cause is to be identified. A thorough history and physical examination of the bird, feather examinations, skin biopsies, blood and faecal tests, X-rays, tests for infectious diseases and allergy testing all will yield useful information. In the short-term, any damaged feathers may need to be plucked, wounds sutured and treated topically and antibiotics given.

The bird should be prevented from inflicting further damage to itself by means of a collar. The exact design will depend upon the areas being plucked or damaged and the bird's individual personality and ability to tolerate the collar whilst still being able to eat and drink. Under no circumstances should a newly collared bird be left unsupervised during the first 24 to 48 hours. The collar should be left on until all wounds have healed and new feathers have grown as both wound healing and feather growth can cause enough irritation for the bird to begin plucking again. Hopefully by this time the underlying causes will have been identified and corrected.

As can be seen there is no magic cure for feather plucking - it requires a thorough work up by the veterinarian and patience and perseverance by the owner for a successful outcome to be reached.

Right:
Mutilated wings of a Sulphur-crested Cockatoo feather plucker.

Below:
Self inflicted wound on the chest of a Galah. The result of feather plucking going too far.

B. DONELEY

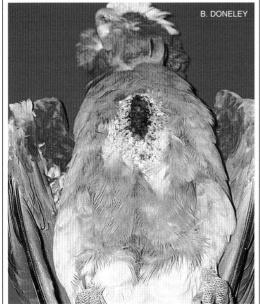

B. DONELEY

Above: Feather plucked Galah.

SPECIES

Introduction

The following section of this book specifically covers each of the six Australian white cockatoo species individually. There is always controversy when discussing subspecies of birds and their distribution. The Australian white cockatoo species is no exception with ornithologists studying these species to determine their true subspecies, family history and distribution. Over the years, many subspecies of these cockatoos have been recognised, and although given, the exact border locations of the areas that they inhabit can be difficult to pinpoint.

It is for this reason that it must be noted that the distribution maps of individual species or subspecies are indicated as accurately as possible, however, this is not to say that these birds do not venture further afield, making it unclear as to the exact extent of their distribution.

It is also for this reason that the subspecies discussed in this book are, at the time of writing, recognised in Australia. This is not to say that other subspecies may exist.

Sulphur-crested Cockatoo

SULPHUR-CRESTED COCKATOO
Cacatua galerita

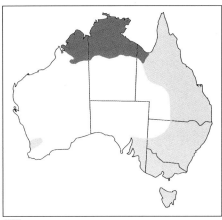

Cacatua galerita galerita

Cacatua galerita fitzroyi

Other Names
White Cockatoo, Greater Sulphur-crested Cockatoo.

Distribution
In the wild, the Sulphur-crested Cockatoo inhabits a large area, taking in northern, eastern, south-eastern Australia, Tasmania, as well as an isolated introduced group inhabiting an area around Perth, Western Australia.

This distribution area features various habitats, including rainforests, eucalyptus bush, open forests, mallee areas, farmland, and in some areas, mangrove swamps.

Species
Of the four subspecies of the Sulphur-crested Cockatoo only two, *Cacatua galerita galerita* and *Cacatua galerita fitzroyi*, inhabit mainland Australia and Tasmania.

Cacatua galerita galerita
C.g. galerita is the nominate subspecies of this group. The length of this cockatoo is around 50cm, weighing approximately 850-900grams. The plumage is primarily white with minimum yellow suffusion around the ear coverts. The undersides of the wings and tail feathers are yellow. The beak is grey-black and the legs dark grey. Distinguishing features are the yellow, forward curving crest and the naked white periophthalmic eye ring.

C.g. galerita is the most common subspecies of Sulphur-crested Cockatoo and can be found throughout Tasmania and Australia, extending eastward from the Gulf of Carpentaria throughout Cape York Peninsula, south through Queensland, New South Wales, Victoria and west as far as Spencers Gulf in South Australia. This subspecies is the introduced population found around Perth, Western Australia.

Cacatua galerita fitzroyi
This subspecies is very similar in appearance to *C.g. galerita*, although much smaller, the average weight being approximately 610-710grams. The yellow ear coverts appear to be more yellow than *C.g. galerita* and the naked periophthalmic eye ring is pale blue in colour. The crest feathers are longer than the nominate subspecies.

C.g. fitzroyi ranges from the Fitzroy River, Western Australia, eastward through northern Australia to the Gulf of Carpentaria where they seem to meet *C.g. galerita*. *C.g. fitzroyi* does not appear to be as common in the wild as *C.g. galerita*, usually only being seen in pairs or small flocks of up to nine birds.

Right; Breeding pair of C.g. galerita Sulphur-crested Cockatoos.

Sexing

Both subspecies of Sulphur-crested Cockatoo are sexually dimorphic, although this is very slight, with the eye colour being the only difference in adult birds.

The cock has dark brown to black eyes, whereas the hen has lighter coloured, sometimes reddish brown eyes. This difference can vary greatly between individual birds making visual sexing by eye colour sometimes difficult.

If purchasing young birds to pair up for future breeding purposes, I would recommend DNA or surgical sexing, to be guaranteed true pairs, as immatures are very difficult to sex.

Breeding

In Australia, the breeding season commences around September, when the cock and hen can be seen working the log.

A clutch of two or three eggs are usually laid at two to three day intervals. Eggs are white, oval in shape and measure approximately 35mm x 28mm. Incubation begins once the first egg is laid and lasts between 25 and 28 days, being shared by both cock and hen. Chicks hatch with a yellow down and are fed by both parents for approximately ten weeks, until fully feathered.

Chicks fledge at around nine to 13 weeks of age. If the parents are tolerating the young, it will not hurt to leave them together for a little longer, as the parents will still feed them for several weeks after they have fledged.

When handraising Sulphur-crested Cockatoos for the pet market I prefer to remove the chicks at approximately three weeks of age as at this stage most of the hard work has already been done by the parents, who will often go back to lay another clutch. This second clutch is usually taken from the parents for handraising around February which concludes the breeding season for these birds.

Above: Sulphur-crested Cockatoo working the nest log.

In captivity, some pairs of Sulphur-crested Cockatoos can take as long as 20 years before they begin to breed, however, once they begin breeding they usually make very good parents and are eager to breed annually thereafter.

I provide a hollow, open-ended, red gum log measuring approximately 1metre long x 30cm internal diameter, fixed to the aviary wall on a slight angle up to 45 degrees, which these

Sulphur-crested Cockatoo chick beginning to hatch.

Sulphur-crested Cockatoo chick, one day old.

Same chick at two days old. Note the thick yellow down.

cockatoos readily accept. Other than hollow logs, metal nestboxes are the only other choice, as wooden nestboxes are unlikely to last a breeding season, due to the birds chewing habits.

During the breeding season some pairs can become very protective and aggressive towards their keeper and although most of the year you can scratch and touch these quiet cockatoos, care should be taken during the breeding season when entering their aviary or carrying out nest inspections. Aggression is also one of the first tell-tale signs that the birds are about to go to nest.

General

These birds are the most well-known of the Australian white cockatoo species. They are very intelligent and strong fliers. The Sulphur-crested Cockatoo has been very popular in Australia as a pet for many years, more so than in aviculture. As this cockatoo seems to adapt to captivity very well, their hardy natures, ability to tame and train to talk and learn tricks, have all been contributing factors to their popularity as a pet.

Many aviculturists overlook these birds to house in aviaries as breeding pairs. However, I feel that due to the years that some pairs seem to take to reach their breeding maturity, any aviculturist who has breeding pairs of Sulphur-crested Cockatoos is very fortunate. Although individual Sulphur-crested Cockatoos, according to most club bird price lists, are very reasonably priced, true breeding pairs demand high prices and are a prized possession.

Above: Flock of wild Sulphur-crested Cockatoos.

In their wild habitat, Sulphur-crested Cockatoos can cause many problems, not only to farmers and their crops, but to country dwellers as well. I remember many years ago when, as an apprentice plumber, I was working on a cedar kit home in Sutton Grange, Victoria where Sulphur-crested Cockatoos are plentiful. Once the carpenters began to install the weatherboards on the house and after they left work each day, the Sulphur-crested Cockatoos in the area would descend on the house and demolish the day's work, chewing and destroying the weatherboards. This cedar house was never completed and was changed to a brick veneer exterior. These birds will also inflict considerable damage to electrical and TV cables on houses if left unprotected. Sulphur-crested Cockatoos are generally described as being noisy, conspicuous birds that probably do not suit the suburban aviculturist. However, they are a great cockatoo to keep and breed and if location permits, I would thoroughly recommend them.

Mutations

There is a report of a Lutino mutation Sulphur-crested Cockatoo being bred in Victoria, however this is the only mutation in this species I am aware of.

Hybrids

Hybrids have been recorded between the Sulphur-crested Cockatoo and other cockatoo species, such as the Short-billed Corella, Eastern Long-billed Corella, Galah and Major Mitchell's Cockatoo.

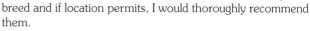

Left: Two Sulphur-crested Cockatoo chicks at 11 weeks of age ready to fledge.

Short-billed Corella

SHORT-BILLED CORELLA
Cacatua sanguinea

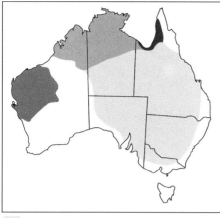

Cacatua sanguinea gymnopis

Cacatua sanguinea sanguinea

Cacatua sanguinea westralensis

Cacatua sanguinea normantoni

Other Names

Little Corella, Bare-eyed Cockatoo, Blood-stained Cockatoo, Blue-eyed Cockatoo.

Distribution

The Short-billed Corella inhabits a large area of Australia, taking in north-western Australia, south-east through central Australia, down the east coast of Australia and through Queensland into New South Wales, Victoria, west through South Australia and mid-western Australia. Short-billed Corellas do not inhabit Tasmania.

Their habitat varies from open forests and scrubland, arid semi-desert, to mangrove swamps. Short-billed Corellas can be seen in very large flocks and never far from a permanent water supply. The Short-billed Corella can also do incredible damage to farming crops.

Species

Until recently there were three recognised subspecies of the Short-billed Corella in Australia. However, Ron Johnstone, Assistant Curator of Ornithology with the Western Australian Museum, believes that there is another subspecies *C.s. westralensis*, located in Western Australia, originally believed to be another group of the *Sanguinea* species.

Cacatua sanguinea sanguinea

C.s sanguinea the nominate subspecies, is approximately 40cm long and weighs approximately 550grams.

Their general plumage is white, with yellow underwings and tail. They have orange or dull red colouring between the eyes and around the bill and have a naked blue-grey periophthalmic eye ring. Legs are grey. Bill is whitish grey in colour. Crest is short and not normally evident. Eyes are dark brown.

C.s. sanguinea is identified from the other Short-billed Corella subspecies by having a paler, blue periophthalmic eye ring and minimal orange or red colouring around the beak.

This species ranges across the north-western area of Australia, west of the Gulf of Carpentaria, toward the Kimberleys and Broome in Western Australia.

Cacatua sanguinea gymnopis

C.s. gymnopis is slightly larger than *C.s. sanguinea* and not quite as white in plumage. The orange or red colouring around the beak appears darker. The naked

Above: *C.s. gymnopis.*

blue-grey periophthalmic eye ring is also darker than *C.s. sanguinea*. Eyes are dark brown.

C.s. gymnopis is the most widespread of the subspecies, extending through the drier inland areas of central, eastern and south-eastern Australia.

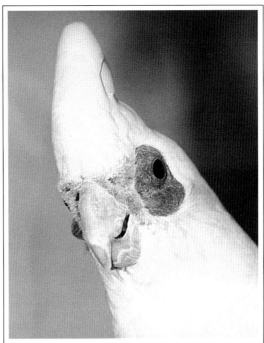

Above: C.s. sanguinea.
Right: C.s. gymnopis.

Above and left: Pair of C.s. gymnopis

Right: Flock of wild corellas in typical habitat.

Cacatua sanguinea normantoni

C.s. normantoni is the smallest of the subspecies and can be identified by having a medium blue-grey coloured periophthalmic eye ring compared with *C.s. sanguinea* and *C.s. gymnopis*. Eyes are dark brown.

Ornithologists are not certain of the exact range that this subspecies inhabit but they do occur around the western Cape York areas, the south-eastern regions of the Gulf of Carpentaria and the southern areas of Papua New Guinea.

Cacatua sanguinea westralensis

Ron Johnstone in the book, *Handbook of Western Australian Birds*, describes members of this subspecies as 'white (underparts often stained or dirty), except for the pale orange lores and more or less concealed orange bases of the head and

Above: Short-billed Corella cock protecting his nest log.

neck feathers; basal portion of the inner webs of primaries and all of the inner webs of the secondaries and the greater underwing coverts are sulphur yellow (forming mostly yellow underwing); and inner webs of all except the central pair of tail feathers are yellow'. The naked periophthalmic eye ring is blue-grey. This subspecies inhabits north and mid-Western Australia.

Sexing

The Short-billed Corella is sexually monomorphic, making sexing very difficult. Surgical or DNA sexing is the only reliable method of determining sex.

Breeding

In Australia, the Short-billed Corella commences breeding around August and continues until the end of December. In captivity, these birds become very noisy during the breeding season.

Usually two to four eggs are laid in a clutch, at two to three day intervals, with incubation being carried out by both parents and lasting approximately 26 days. Eggs are white, oval in shape and measure approximately 41mm x 29mm.

Chicks hatch covered in yellow down and remain in the nest for between seven and eight weeks. The young are fed by both parents while in the nest and for several weeks after leaving the nest.

I have found that if the young are taken for handraising at around two to three weeks of age, the parents will usually double brood.

I have very good success breeding Short-billed Corellas in hollow logs, measuring 1 metre long x 30cm internal diameter, hung on a slight angle up to 45 degrees in the aviary.

Whilst working their log they will completely remove the nesting material therefore it is very important to observe the amount remaining within the log and add more daily, if necessary.

Left: Short-billed Corellas at metal nest.

General

The Short-billed Corella is reportedly the first Australian parrot to be sighted by explorers in this country. They are nomadic and can be seen in flocks of thousands of birds, especially at the end of the breeding season, feeding in paddocks during early morning and late afternoon. These birds are never found very far from water. In fact, a flock of Short-billed Corellas was a very good guide to the Australian aborigines that water could be found nearby to that area. In the wild, these birds are the most widespread of all the corella species. They are known as noisy, conspicuous birds with flocks being heard from afar.

Short-billed Corellas are very hardy birds and when handraised, they make excellent pets and are extremely good talkers. This little corella thrives in captivity, is easily catered for, easy to breed and therefore, an ideal cockatoo species for the beginner or novice.

Unfortunately, because this bird is relatively common, it is not very popular with aviculturists. Once again these cockatoos can become very noisy and are not recommended to be kept in suburban areas. However, they are good 'watch birds' and will attract your attention to anything or anyone strange around your premises.

Personally, I regard Short-billed Corellas as lovely birds and enjoy them in my collection.

Mutations

There are no mutations of the Short-billed Corella recorded.

Hybrids

Hybrids have been recorded between Sulphur-crested Cockatoos, Major Mitchell's Cockatoos and Galahs.

Right: Short-billed Corella chicks - six weeks of age.

Above
Short-billed Corella chicks - two weeks of age.
Below:
Short-billed Corella chicks - four weeks of age.

Eastern Long-billed Corella

EASTERN LONG-BILLED CORELLA
Cacatua tenuirostris

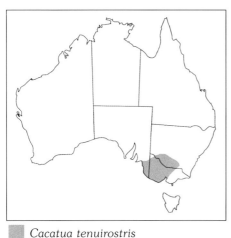

Cacatua tenuirostris

Other Names
Slender-billed Cockatoo, Long-billed Cockatoo.

Distribution
The Eastern Long-billed Corella ranges throughout south-eastern Australia including western Victoria, south-western New South Wales, and south-eastern South Australia.

This distribution area includes habitat such as open forests, farmlands, woodlands and grasslands. They are generally never found very far from water.

Although these cockatoos can be seen in some areas congregating in large flocks, most reports as to the habitat and status in the wild show a definite decline in this species and its range.

Species
It was once considered that the Western Long-billed Corella was a subspecies of the Eastern Long-billed Corella, being named *Cacatua tenuirostris pastinator*. However, there have been no subspecies of *C. tenuirostris* recorded or confirmed.

Cacatua tenuirostris
The Eastern Long-billed Corella is approximately 40cm long, weighing approximately 640grams. The cock appears to be heavier than the hen.

This cockatoo can be identified by its generally white plumage with faint yellow on the underwings and undertail. An orange or reddish band runs across the throat, forehead, nape, lores, mantle and breast. As in other corellas, they have a naked periophthalmic eye ring, blue-grey in colour. Eyes are dark brown. Legs are grey. The characteristic long, slender bill, a whitish grey colour, is adapted for digging roots and bulbs as a food supply. As with all corellas the crest is short and when erected, is helmet-like in appearance.

Sexing
Being monomorphic the Eastern Long-billed Corella is extremely difficult to sex. Both adults and young birds require surgical or DNA sexing to establish a true pair.

Breeding
In Australia, the breeding season of the Eastern Long-billed Corella commences around late August continuing through to December.

The Western Long-billed Corella was once considered a subspecies of the Eastern Long-billed Corella. As seen in this comparison the Eastern Long-billed Corella (left) features a red throat and longer, slimmer bill. The Western Long-billed Corella is a larger and heavier species.

Two or three, sometimes four eggs will be laid at two to three day intervals. Eggs are white, long-oval in shape and measure approximately 51mm x 30mm. Incubation can commence after the first egg is laid, however, usually incubation will begin with the laying of the second egg. Incubation is carried out by both parents taking shifts, normally the cock during daylight hours and the hen during the night. The incubation period of this corella species is around 24 days, although longer periods have been recorded.

Chicks are born covered with yellow down and remain in the nest for six to eight weeks, being fed by both parents. Once fledged, the parents will continue to feed the young for several weeks until about 12 weeks of age, when they are independent.

Once again I have found that an open-ended red gum log measuring approximately 1 metre long x 30cm internal diameter, hung on a slight angle up to 45 degrees, serves the breeding purpose of these corellas satisfactorily.

This species can prove to be more difficult to breed than some other cockatoo species.

Above: Pair of Eastern Long-billed Corellas.
Left: Eastern Long-billed Corella chick approximately three weeks of age.

General

The Eastern Long-billed Corella is reasonably uncommon in captivity throughout Australia.

These birds can be very pugnacious towards other birds and for this reason they are best housed as individual pairs.

Concrete floors are recommended as their incredible ability to dig, tunnel and burrow, could soon see them out of an earthen floored aviary. Many years ago, a friend purchased a pair of Eastern Long-billed Corellas and placed them in a bank of aviaries, their enclosure measuring 2 metres square. A week later I was invited around to view these birds. To the owner's horror, they were gone! They had tunnelled about 3 metres (luckily towards another aviary) and were sitting in the next aviary alongside a pair of Alexandrine Parrots. Stories abound when discussing these birds, in fact, there are reports of Eastern Long-billed Corellas laying eggs, incubating and even raising young in burrows when nesting facilities have been unavailable or unacceptable to them.

This species' digging habits possibly aids the trimming and well-being of their long, slender bill, and also combats boredom. I provide a tray full of dirt and sand in their aviary to allow this practice.

When taken and handreared these birds make wonderful pets and are extremely good talkers. They can be extremely noisy and are better kept in collections away from suburban areas.

Mutations

There are no known mutations in the Eastern Long-billed Corella.

Hybrids

Hybrids occur with Sulphur-crested Cockatoos, Major Mitchell's Cockatoos and Galahs.

C.p. pastinator

Western Long-billed Corella

Cacatua pastinator

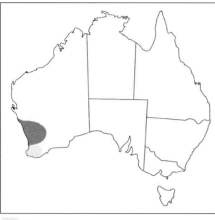

Cacatua pastinator pastinator

Cacatua pastinator butleri

Other Names
Dampier's Cockatoo.

Distribution
The Western Long-billed Corella is restricted to the south-western area of Western Australia, usually seen in small flocks, although when roosting, they can gather in large flocks of 1,000 or more birds.

They inhabit areas of farmland, open forest and water courses in timber lands.

Species
There are two recognised subspecies of Western Long-billed Corella in Australia. These can be seen in two distinctly different areas within their range.

Cacatua pastinator pastinator
C.p. pastinator, the nominate subspecies, is approximately 43-48cm long and weighs approximately 650grams. This subspecies can be identified by being predominately white with the underparts often dirty or stained. Lores are orange. Underwing coverts and undertail are sulphur yellow. The crest of these birds is short. Eyes are dark brown. Naked periophthalmic eye ring is bluish grey. The bill is greyish white in colour and the legs are dark grey. The heavier bill of this bird is believed to have evolved as a result of digging in heavier soils.

C.p. pastinator ranges throughout a small area of the south-western inland region near Boyup Brook and Qualeup, south to the Perup River, Lake Muir and Cambellup, Western Australia.

In the wild, this subspecies mainly inhabits farmlands and is unevenly distributed throughout other areas of its range.

Above: Young pair *C.p. pastinator*.

It is considered to be the rarest cockatoo in Australia, with numbers in the wild estimated to be as low as 1600 birds. Due to this declining number, a captive breeding program was approved and a total of ten nestlings were taken from nests in the wild over a period of three years by Conservation and Land Management (CALM) in Western Australia and raised at the Perth Zoo. This was Stage One of the program. Stage Two will involve further nestlings being taken from nests in the wild and placed in the care of several approved Western Australian aviculturists. The aim of this program is to establish a captive gene pool in case there is a further decline in the wild.

Cacatua pastinator butleri
C.p. butleri is very similar in colour and appearance to *C.p. pastinator*, except it is generally smaller, with less red around the smaller bill.

This subspecies inhabits lightly timbered country as well as farmlands well north of the *C.p. pastinator* and ranges from around Dongara, south around Jurien across to Moora, Northam, Brookton and east to the upper Moore River, Victorian Plains, Wongan Hills and Koorda, Western Australia.

Ref. *Handbook of Western Australian Birds.*

Above: *C.p. butleri* juvenile group.

Sexing

The Western Long-billed Corella is sexually monomorphic and extremely difficult to sex although cocks do tend to be larger in the head, however, this is not always true. If true pairs are to be obtained, surgical or DNA sexing are the only reliable methods.

Breeding

In Australia, the breeding season of the Western Long-billed Corella usually commences around September to the end of November.

One to three eggs are laid at two to four day intervals. Eggs are similar to those of the Eastern Long-billed Corella, being white, long-oval in shape and approximately 51mm x 30mm. Incubation can begin with the laying of the first egg but generally after the second, lasting 26 to 28 days.

The chicks hatch once again in true cockatoo fashion, covered in yellow down, although their long, slender bill is evident.

The chicks are fed by both parents whilst in the nest and for several weeks once fledged. They fledge between six to eight weeks of age. Young can be removed from their parents around ten weeks of age, however if any aggression is shown from either parent, the chicks should be removed immediately.

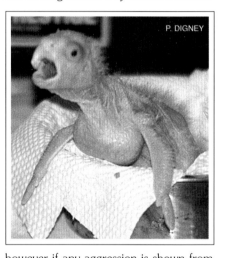

P. DIGNEY

Hollow logs or metal nesting receptacles of similar size to those previously mentioned, are recommended as nesting sites.

General

As mentioned previously, the Western Long-billed Corella was once considered a subspecies of the Eastern Long-billed Corella and known as *Cacatua tenuirostris pastinator*. Over the years much study and work has been conducted, not only on this species but all species of corella, with the Western Long-billed Corella being classified as a specific species, *Cacatua pastinator pastinator*. These studies have

Above: *Handreared Western Long-billed Corella chick approximately three weeks of age.*
Left: *C.p. butleri pair.*

shown that the Western Long-billed Corella has a totally different temperament to that of the other corella species.

Many of these cockatoos have been kept in their native state of Western Australia for many years, as they make excellent handreared pets.

The Western Long-billed Corella is one of the corella species which Western Australian aviculturists are beginning to take seriously, keeping and breeding them in their aviaries. Many of these birds have been exported, over the last two to three years, to the eastern states of Australia.

If contemplating keeping a pair of these birds, they are very hardy, easily catered for and devoted pairs will breed quite readily.

They can be very noisy and are therefore best kept away from suburban areas. The antics of these birds will keep you entertained for hours which I am sure is a contributing factor to their popularity as pets and aviary subjects. Western Long-billed Corellas do make excellent pets and are one of the best talkers in the white cockatoo family.

Mutations
There has been no documented evidence to suggest any mutations.

Hybrids
No hybrids of the Western Long-billed Corella are recorded. However, in the wild, any hybridisation between the Western Long-billed Corella and the Short-billed Corella that may exist would be very hard to depict and would possibly be undetected.

Right:
Juvenile C.p. pastinator.

Left: C.p. pastinator juvenile cock.
Note large lower mandible adapted
for digging in heavy soil.

Major Mitchell's Cockatoo

MAJOR MITCHELL'S COCKATOO
Cacatua leadbeateri

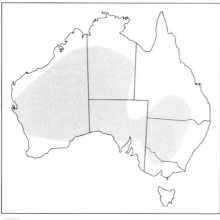

Cacatua leadbeateri leadbeateri

Other Names
Pink Cockatoo, Leadbeater's Cockatoo, Wee Juggler, Cocklerina, Chockalott.

Distribution
The Major Mitchell's Cockatoo inhabits inland Australia, covering areas of Queensland, New South Wales, Victoria, South Australia, Northern Territory and Western Australia.

It is common in areas around the Great Australian Bight although at times irregularly distributed, and usually seen in pairs or small flocks in other areas throughout its range.

The Major Mitchell's Cockatoo inhabits dry scrubland, open woodlands and farmlands and arid, harsh grasslands, never far from a water supply.

Species
There are two recorded subspecies of Major Mitchell's Cockatoo in Australia, *Cacatua leadbeateri leadbeateri* and *Cacatua leadbeateri mollis*. Some study in Western Australia has led some ornithologists to believe the colour variations of the crests of these subspecies point to the distribution areas of these birds in the wild, rather than them being subspecies. In other areas of Australia groups of these birds can vary greatly in colour with some featuring less salmon pink colouring than other Major Mitchell's Cockatoos. I believe a lot more study and time will have to be devoted to this subject to determine an explanation for these occurrences.

Cacatua leadbeateri leadbeateri
C.l. leadbeateri is the nominate subspecies, measuring approximately 39cm in length and weighing approximately 420grams.

The crown is white suffused with salmon pink colouring, with a forward curving crest of red or scarlet feathering with a central band of yellow. The forehead and sides of the head are red or scarlet. Nape, breast, belly and underwings are salmon pink. Beak is bone coloured. Legs are grey. Naked periophthalmic eye ring is greyish white in colour.

The distribution of this subspecies ranges over the eastern, central, southern and western states of Australia. *C.l. leadbeateri* is the most common subspecies in both the wild and in captivity.

Cacatua leadbeateri mollis
This subspecies can be identified from *C.l. leadbeateri* by having very little if any yellow crest feathers, making the crest appear to be dark red. This is particularly noticeable in the Kimberley populated area. *C.l. mollis* inhabits the south-western region of Australia, although the exact extent of its range is unclear.

In some breeding examples of the *C. l. mollis* the progeny have shown yellow in the crest. One

Right: C.l. leadbeateri pair.

explanation for this could be that the cross breeding of *C.l. leadbeateri* and *C.l. mollis* in the earlier years of aviculture, resulted in some red-crested birds still carrying yellow-crested genes through to today's generations.

Sexing

Being dimorphic, there are two ways to determine the sex of Major Mitchell's Cockatoos.

Sex can be determined by the crest. The feathers in the crest of the cock are closer together, whereas in the hen there is a slight separation between each feather. The hen tends to have a wider yellow band in the crest while the cock has a narrower, brighter and more uniformed yellow crest band, although this is not applicable in *C.l. mollis*.

The eye colour can also determine the sex of Major Mitchell's Cockatoos. The eye colour of the cock is dark brown or black, whereas the hen has a reddish light brown eye colour. In young chicks, the eye colour of the cock and hen is the same, changing around eight months of age.

D. ANDERSEN

D. ANDERSEN

Above: C.l. mollis cock.
Left: C.l. mollis hen.

Breeding

In Australia, Major Mitchell's Cockatoos usually go to nest around September or October.

Two, sometimes three eggs are laid at two to three day intervals. Eggs are white, oval in shape and measure approximately 38mm x 28mm. Incubation generally commences with the laying of the second egg, and is carried out by both parents for a period of 24 to 26 days.

Chicks are born almost naked and fledge between seven and eight weeks of age. The young fledglings should remain with their parents for a further six to eight weeks to ensure that they are fully weaned and self-sufficient before being removed.

I provide a vertically hung hollow log 1 metre long x 30cm internal diameter with an entrance hole near the top of the log approximately 16cm in diameter.

Major Mitchell's Cockatoos have a reputation for becoming aggressive towards their young fledglings. Therefore, it is advisable to keep a strict eye on both the parents and the young during this time, until the young are old enough to be removed. It should be stressed however, that at the first sign of aggression, the young birds must be separated from the parents immediately.

Major Mitchell's Cockatoos are one of the easiest cockatoo species to breed.

General

The Major Mitchell's Cockatoo is possibly the most commonly kept and bred cockatoo in Australian aviculture. These birds can be noisy especially at first light and dusk which does not make them a suitable species for suburban backyards.

This species, especially young birds, can be very destructive, therefore it is advisable that they are provided fresh eucalyptus branches to chew and destroy daily. They seem to do better in spacious aviaries, however, their needs are basic and normally they will become quite tame towards their owners.

Handreared birds do not usually make good pets. Although they can become as quiet as other handreared cockatoo species, I believe they are too shy and nervous to make good pets. Handreared birds, when adults, can become very aggressive toward their owners, especially during nest inspections. Due to their aggression, if contemplating keeping and breeding Major Mitchell's Cockatoos, they

Above: C.l. mollis cock, note lack of yellow feathering in crest.
Left: C.l. leadbeateri hen showing distinctive yellow band and separated feathers in crest.

should be housed as individual pairs, and in this way I am sure that you will find them a pleasure to keep.

Mutations
There are no recorded mutations in the Major Mitchell's Cockatoo.

Hybrids
Hybrids have been recorded with Sulphur-crested Cockatoos, Short-billed Corellas and Galahs.

Left: C.l. leadbeateri. The cock in the foreground features a narrower, brighter and more uniformed yellow crest band.

Right: C.l. leadbeateri pair at their nest log.

D. ANDERSEN

Above:
Major Mitchell's chicks at 10-13 days old.
Right:
Major Mitchell's chick at 11 days old.
Below:
Major Mitchell's chicks at 28-32 days old.

D. ANDERSEN

D. ANDERSEN

D. ANDERSEN

Above: Major Mitchell's chicks at 42-45 days old.

C.r. roseicapilla.

Galah

GALAH
Cacatua roseicapilla

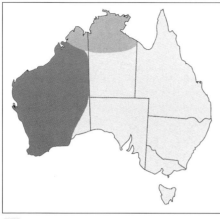

☐ *Cacatua roseicapilla roseicapilla*

▨ *Cacatua roseicapilla kuhli*

■ *Cacatua roseicapilla assimilis*

Other Names
Roseate Cockatoo, Rose-breasted Cockatoo.

Distribution
Galahs range throughout Australia. *C.r. roseicapilla* is an introduced species in Tasmania.

The Galah is the most commonly found cockatoo in Australia occupying a wide variety of habitat including eucalyptus bush, open forests and farmlands.

Species
There are three recognised subspecies of Galahs inhabiting Australia.

Cacatua roseicapilla roseicapilla
(Eastern subspecies)

C.r. roseicapilla is the nominate subspecies measuring approximately 35cm in length and weighing approximately 340grams. Underparts and underwing coverts are rose pink with upper parts a soft, dove grey colour. Beak is a whitish colour

Above: C.r. roseicapilla.
Left: Young C.r. roseicapilla hen.

and legs are grey.

It can be identified from the other two subspecies by having a whiter crest which has a definite break in the crest feather formation when erect and a pink periophthalmic ring around the dark brown-black eye. However, I have noted, looking at the different areas that the eastern subspecies Galah inhabits ie Queensland, New South Wales, Victoria, Central Australia and South Australia, that the naked periophthalmic eye ring colour can vary from dark pink to light pink, which I am sure with more study could possibly pinpoint the region that a particular pair of birds originate.

Above: C.r. assimilis hen. Crest is fuller and deeper pink in colour. The periophalmic eye ring is almost white in colour.

Above: C.r. roseicapilla. Crest is whiter with a definite break in feather formation. The periophthalmic eye ring is a pink colour

Distribution ranges over the eastern, north-eastern, southern and central Australian areas as well as Tasmania.

Cacatua roseicapilla assimilis
(Western subspecies)

C.r. assimilis, the western subspecies, is identified by having a longer, fuller and deeper pink crest with no division in the feather formation compared with the eastern subspecies. The naked periophthalmic eye ring is almost white. Eye is brown.

C.r. assimilis tends to be longer and slightly broader than C.r. roseicapilla with a paler body colouration. The head is bigger than its eastern counterpart and appears flatter and more square looking.

These birds normally range in the west from around Derby in the north-west, east across the border of Western Australia and south including all of south-western Australia.

Above: Pair C.r. assimilis.
Left: Pair C.r. roseicapilla.

Cacatua roseicapilla kuhli
(Northern subspecies)

C.r. kuhli is considerably smaller and lighter in colour than the other two subspecies.

The head of *C.r. kuhli* is smaller, with a pink crest, and the naked periophthalmic eye ring is more reddish in colour than the other subspecies.

Distribution ranges north of Derby, north-west Western Australia and north throughout the Northern Territory.

Sexing

Galahs are dimorphic and easily sexed by eye colour once they have reached maturity at approximately 12 months of age. The cock has a dark brown-black eye whereas the hen has a light copper-red coloured eye.

Breeding

In captivity, they are usually the first birds in a cockatoo collection to begin nesting, laying from as early as July-August through to December.

C.r. kuhli is the smallest of the subspecies showing the periophthalmic eye ring more reddish in colour.

They usually lay three to four eggs at two to three day intervals. Eggs are white, oval in shape and measure approximately 35mm x 26mm. The eggs are incubated by both parents for 23 to 25 days, commencing after the last egg is laid. Chicks hatch covered in pinkish-white down and are fed by both parents. They fledge at six to seven weeks of age.

In the wild, the Galah does not double brood and they don't normally breed until they are two to three years of age. In captivity, however, they may go to nest again if you remove the young from the first clutch for handraising at two to three weeks of age.

Galahs nest in eucalyptus trees, therefore, it is preferable that a log be provided when breeding in captivity is planned. However, if a log is not available a nestbox will suffice.

Galahs have two unusual habits when they begin to nest. Firstly, they take eucalyptus leaves, branches and twigs into their nests, to chew up and lay their eggs

Above and left: C.r. kuhli.

on. When breeding the Galah in captivity, it is very important to provide fresh eucalyptus branches daily from the time the birds begin to work their log right up until when they lay. When they leave the nest they will cover their eggs with a few leaves. I provide fresh branches in branch holders situated alongside the nest entrance for this reason.

Secondly, Galahs will strip and chew bark from around their nesting hollow entrance until it is almost polished around the entire circumference of the hollow. This behaviour is said to be practised to protect themselves, eggs and young from predators, therefore making it slippery and very hard, or impossible, to enter the nest. Both parents participate in this activity.

In the wild they will visit the nest many times throughout the year to maintain ownership of the site, likewise, in captivity they can be seen working their logs at various times throughout the year.

Breeding pairs are closely bonded and will spend most of the day together sitting, or mutually preening each other. Even when incubating, the other partner is normally close by or at the nest entrance.

In captivity, different sized logs can be used from about 50cm to 2 metres long with an internal diameter of 20cm to 30cm. These can be hung either vertically or horizontally. If a nestbox is used for breeding it is wise to make it approximately 30cm square x 90cm deep, with a closed top and bottom, and a natural spout on the side. Due to the difficulty of obtaining logs large enough for breeding cockatoos, this design can also be used for other cockatoo species. A mixture of sawdust and peatmoss is used as nesting materials in the bottom of the log or box to a depth of 10cm to 12cm.

General

Although the Galah is common and inexpensive it is a great little cockatoo, abundant in character, and will breed quite readily in an aviary situation. If you purchase any of the other cockatoo species, such as Gang Gangs, Major Mitchell's or Black Cockatoos, you may have to wait for up to five years, or more, before they breed, whereas by starting off with a young pair of Galahs they could commence breeding when they are two to three years of age. They are the perfect cockatoo to gain experience with before moving to the more temperamental and more difficult to breed species.

Galahs can live as long as humans and are capable of breeding for up to 40 years. A breeding pair of Galahs are capable of breeding 120 young in their lifetime.

Top:
Galah chicks at 17 days old.
Note Lutino on right.
Centre:
Galah chicks at 24 days old.
Bottom:
Galah chicks at 31 days old.
Note the nearly fully feathered Lutino chick on right.

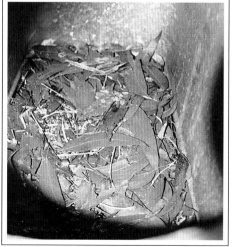

Above: Galah nest lined with gum leaves.
Left: C.r. roseicapilla at nest.

Young Galahs are eagerly sought after as pets, making handreared birds very popular. In fact, most people wanting to purchase these birds as pets must order their birds before the breeding season begins, to ensure that they don't miss out.

They become very quiet and tame towards their owners, will talk well and are very hardy birds.

They will eat almost anything that is offered to them, however a strict watch on diet is essential as Galahs are prone to obesity. Therefore, Budgerigar seed, and no sunflower seed, is recommended as the only dry seed mix fed permanently in an aviary situation.

There is however, one downfall in keeping and breeding Galahs, and that is, as in the Gang Gang Cockatoo, some birds are prone to feather plucking, which I believe is related to boredom or stress. They can also become very excited and this leads to them screeching and whistling very loudly, so consideration once again should be given to neighbours especially if you live in a residential area.

I continue to be amazed at the Galah's lack of popularity among Australian aviculturists as it is a most interesting cockatoo that is well suited to aviary life and to both novice and experienced breeders alike.

Mutations

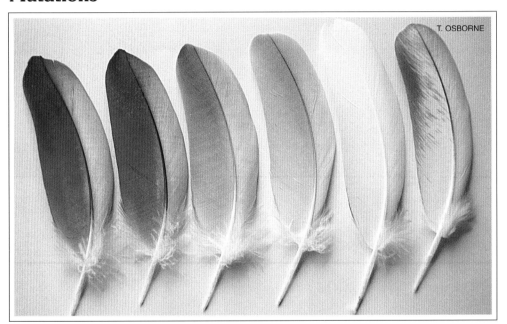

T. OSBORNE

Above: Feathers from left to right: Normal, Blue, Isabel, Cinnamon, Lutino and Pied.

Black-eyed White
(Recessive Inheritance)

This bird has a pink breast, white wings and greyish tones in the tail and flight feathers. Eyes are black. This is not a good name for this mutation. However, combining it with a Blue mutation will produce a true Black-eyed White. It is actually the equivalent of the Yellow mutation in a green parrot. Perhaps Pink is a more appropriate name.

Left: Black-eyed White Galah mutation.

Above: Blue Galah mutation.
Right: Lutino Galah mutation.

Left: Cinnamon Galah mutation.
Below: Pied Galah mutation.
Bottom left: This is a new Galah mutation, possibly an Isabel which is yet to be established.

Blue (Grey and White) (Recessive Inheritance)

Pink is replaced with white on this bird. This a typical 'Blue' mutation, where all psittacin pigments are removed leaving melanins unaffected.

Lutino (Sex-linked Inheritance)

The breast and abdomen are pink with wings being pure white. Eyes are red and feet are pink This is a typical 'Lutino' mutation where all melanin pigments are removed and psittacin pigments are unaffected.

Cinnamon (Sex-linked Recessive Inheritance)

This bird has cinnamon flight feathers, its eyes are dark and feet are pink. The Cinnamon mutation converts black and grey into brown melanin, an effect that is clearly visible.

Silver (Recessive Inheritance)

The normal grey wings are diluted to silver. It is a typical Dilute mutation, correctly called Silver (dilute of normal) in a grey coloured bird. It has been incorrectly referred to as a recessive Cinnamon.

Pied

Breast and abdomen are pink with pied markings in the wings and flight feathers.

Hybrids

Hybridisation has been recorded with Sulphur-crested, Major Mitchell's and Gang Gang Cockatoos, and Short and Long-billed Corellas.

LUTINO
Sex-linked Recessive

COCK		HEN
Normal/Lutino	x	**Normal**
Normal		Normal
Normal/Lutino		Lutino
Lutino	x	**Normal**
Normal/Lutino		Lutino
Normal	x	**Lutino**
Normal/Lutino		Normal
Normal/Lutino	x	**Lutino**
Normal/Lutino		Normal
Lutino		Lutino
Lutino	x	**Lutino**
Lutino		Lutino

CINNAMON
Sex-linked Recessive

COCK		HEN
Normal/Cinnamon	x	**Normal**
Normal		Normal
Normal/Cinnamon		Cinnamon
Cinnamon	x	**Normal**
Normal/Cinnamon		Cinnamon
Normal	x	**Cinnamon**
Normal/Cinnamon		Normal
Normal/Cinnamon	x	**Cinnamon**
Normal/Cinnamon		Normal
Cinnamon		Cinnamon
Cinnamon	x	**Cinnamon**
Cinnamon		Cinnamon

SILVER
Autosomal Recessive
For recessive mutations, the relative sex of each parent is irrelevant and all outcomes occur   in both sexes.

Normal/Silver	x	**Normal**
Normal and Normal/Silver cocks and hens		
Normal	x	**Silver**
Normal/Silver cocks and hens		
Normal/Silver	x	**Silver**
Normal/Silver and Silver cocks and hens		
Silver	x	**Silver**
Silver cocks and hens		

BLUE (Grey & White)
Autosomal Recessive
For recessive mutations, the relative sex of each parent is irrelevant and all outcomes occur in both sexes.

Normal/Blue	x	**Normal**
Normal and Normal/Blue cocks and hens		
Normal	x	**Blue**
Normal/Blue cocks and hens		
Normal/Blue	x	**Blue**
Normal/Blue and Blue cocks and hens		
Blue	x	**Blue**
Blue cocks and hens		

BLACK-EYED WHITE (BEW)
Autosomal Recessive
For recessive mutations, the relative sex of each parent is irrelevant and all outcomes occur in both sexes.

Normal/BEW	x	**Normal**
Normal and Normal/BEW cocks and hens		
Normal	x	**BEW**
Normal/BEW cocks and hens		
Normal/BEW	x	**BEW**
Normal/BEW and BEW cocks and hens		
BEW	x	**BEW**
Black-eyed White cocks and hens		

COMBINATIONS OF MUTATIONS – the above mutations can be combined to give the following combinations. Few if any of these combinations have been produced as yet, so their actual appearance is theoretical and is extrapolated from knowledge of the same mutations in other species of parrots.

ALBINO
Combination of Lutino and Blue

COCK		HEN
Lutino	x	**Blue**
Normal/Lutino/Blue		Lutino/Blue
Blue	x	**Lutino**
Normal/Lutino/Blue		Normal/Blue
Normal/Lutino/Blue	x	**Lutino**
Normal/Lutino		Normal
Normal/Lutino/Blue		Normal/Blue
Lutino		Lutino
Lutino/Blue		Lutino/Blue
Normal/Lutino/Blue	x	**Blue**
Normal/Blue		Normal/Blue
Normal/Lutino/Blue		Lutino/Blue
Blue		Blue
Blue/Lutino		Albino
Blue/Lutino	x	**Lutino/Blue**
Normal/Lutino/Blue		Normal/Blue
Lutino/Blue		Lutino/Blue
Blue/Lutino		Blue

Albino		Albino
Lutino/Blue	x	**Blue**
Normal/Lutino/Blue		Lutino/Blue
Blue/Lutino		Albino
Blue/Lutino	x	**Albino**
Blue/Lutino		Blue
Albino		Albino
Lutino/Blue	x	**Albino**
Lutino/Blue		Lutino/Blue
Albino		Albino
Albino	x	**Albino**
Albino		Albino

IVORY

Cinnamon Blue combination – could be called Fawn in a grey-based species.

COCK		**HEN**
Cinnamon	x	**Blue**
Normal/Cinnamon/Blue		Cinnamon/Blue
Blue	x	**Cinnamon**
Normal/Cinnamon/Blue		Normal/Blue
Normal/Cinnamon/Blue	x	**Blue**
Normal/Blue		Normal/Blue
Normal/Cinnamon/Blue		Cinnamon/Blue
Blue		Blue
Blue/Cinnamon		Ivory
Blue/Cinnamon	x	**Blue**
Blue		Blue
Blue/Cinnamon		Ivory
Blue/Cinnamon	x	**Cinnamon/Blue**
Normal/Cinnamon/Blue		Normal/Blue
Blue/Cinnamon		Blue
Cinnamon/Blue		Cinnamon/Blue
Ivory		Ivory
Blue/Cinnamon	x	**Ivory**
Blue/Cinnamon		Blue
Ivory		Ivory
Cinnamon/Blue	x	**Blue**
Normal/Cinnamon/Blue		Cinnamon/Blue
Blue/Cinnamon		Ivory
Cinnamon/Blue	x	**Ivory**
Cinnamon/Blue		Cinnamon/Blue
Ivory		Ivory
Ivory	x	**Ivory**
Ivory		Ivory

SILVER CINNAMON

In this combination, diluting cinnamon creates cream coloured wings with pink breast.

COCK		**HEN**
Cinnamon	x	**Silver**
Normal/Cinnamon/Silver		Cinnamon/Silver
Normal/Cinnamon/Silver	x	**Silver**
Normal/Silver		Normal/Silver
Normal/Cinnamon/Silver		Cinnamon/Silver
Silver		Silver

Silver/Cinnamon		Silver Cinnamon
Silver/Cinnamon	x	**Silver**
Silver		Silver
Silver/Cinnamon		Silver Cinnamon
Silver/Cinnamon	x	**Cinnamon/Silver**
Normal/Cinnamon/Silver		Normal/Silver
Cinnamon/Silver		Silver
Silver/Cinnamon		Cinnamon/Silver
Silver Cinnamon		Silver Cinnamon
Silver/Cinnamon	x	**Silver Cinnamon**
Silver/Cinnamon		Silver
Silver Cinnamon		Silver Cinnamon
Cinnamon/Silver	x	**Silver Cinnamon**
Cinnamon/Silver		Cinnamon/Silver
Silver Cinnamon		Silver Cinnamon
Silver Cinnamon	x	**Silver Cinnamon**
Silver Cinnamon		Silver Cinnamon

SILVER BLUE

For combinations of two recessive mutations, the relative sex of each parent is irrelevant and all outcomes occur in both sexes.

Silver	x	**Blue**
Normal/Silver/Blue cocks and hens		
Normal/Silver/Blue	x	**Normal/Silver/Blue**
Normal, Normal/Blue, Normal/Silver, Normal/Silver/Blue, Blue, Blue/Silver, Silver, Silver/Blue, Silver Blue cocks and hens		
Blue/Silver	x	**Silver/Blue**
Normal/Silver/Blue, Blue/Silver, Silver/Blue, Silver Blue cocks and hens		
Blue/Silver	x	**Silver Blue**
Blue/Silver and Silver Blue cocks and hens		
Silver/Blue	x	**Silver Blue**
Silver/Blue and Silver Blue cocks and hens		
Silver Blue	x	**Silver Blue**
100% Silver Blue cocks and hens		

CREAM (SILVER IVORY)

A combination of Silver, Cinnamon and Blue.

COCK		HEN
Ivory	x	**Silver Blue**
Blue/Cinnamon/Silver		Ivory/Silver
Silver Blue	x	**Ivory**
Blue/Cinnamon/Silver		Blue/Silver
Silver Cinnamon	x	**Silver Blue**
Silver/Cinnamon/Blue		Silver Cinnamon/Blue
Silver Blue	x	**Ivory/Silver**
Blue/Cinnamon/Silver		Blue/Silver
Silver Blue/Cinnamon		Silver Blue
Silver Cinnamon	x	**Ivory/Silver**
Cinnamon/Blue/Silver		Cinnamon/Blue/Silver
Silver Cinnamon/Blue		Silver Cinnamon/Blue
Blue/Cinnamon/Silver	x	**Silver Blue**
Blue/Silver		Blue/Silver
Silver Blue		Silver Blue

Blue/Cinnamon/Silver   Ivory/Silver
Silver Blue/Cinnamon   Cream
Silver Blue/Cinnamon   x   **Ivory/Silver**
Blue/Cinnamon/Silver   Blue/Silver
Ivory/Silver   Ivory/Silver
Silver Blue/Cinnamon   Silver Blue
Cream   Cream
Silver Blue/Cinnamon   x   **Cream**
Silver Blue/Cinnamon   Silver Blue
Cream   Cream
Silver Cinnamon/Blue   x   **Cream**
Silver Cinnamon/Blue   Silver Cinnamon/Blue
Cream   Cream
Cream   x   **Ivory/Silver**
Ivory/Silver   Ivory/Silver
Cream   Cream
Cream   x   **Silver Cinnamon/Blue**
Silver Cinnamon/Blue   Silver Cinnamon/Blue
Cream   Cream
Cream   x   **Cream**
Cream   Cream

TRUE WHITE

Combination of Black-eyed White (BEW) and Blue – this combination will be pure white with a black *eye*.

For combinations of two recessive mutations, the relative sex of each parent is irrelevant and all outcomes occur in both sexes.

BEW   x   **Blue**
Normal/Blue/BEW cocks and hens
Normal/Blue/BEW   x   **Normal/Blue/BEW**
Normal, Normal/Blue, Normal/BEW, Normal/Blue/BEW, Blue, Blue/BEW, BEW, BEW/Blue and White cocks and hens
BEW/Blue   x   **Blue/BEW**
Normal/Blue/BEW, Blue/BEW, BEW/Blue and White cocks and hens
BEW/Blue   x   **White**
BEW/Blue and White cocks and hens
Blue/BEW   x   **White**
Blue/BEW and White cocks and hens
White   x   **White**
100% White cocks and hens

REFERENCES

Handbook of Western Australian Birds (Vol. 1)
by R E Johnstone & G M Storr - 1998 Western Australian Museum.
Cockatoos In Aviculture,
by Rosemary Low - 1993 Blandford Books.
Australian Cockatoos
by Stan Sindel & Robert Lynn - 1989 Singil Press.
A Guide to Basic Health and Disease in Birds,
by Dr Michael Cannon BVSc MACVSc (Avian Health) - 1996 ABK Publications.
A Guide to Incubation and Handraising Parrots
by Phil Digney - 1998 ABK Publications.

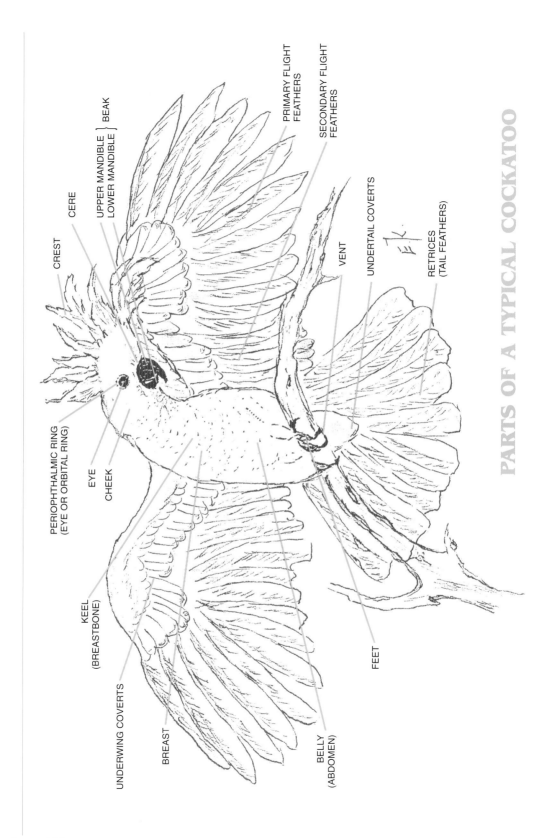

PARTS OF A TYPICAL COCKATOO

CREST

CERE

UPPER MANDIBLE } BEAK
LOWER MANDIBLE

PRIMARY FLIGHT FEATHERS

SECONDARY FLIGHT FEATHERS

UNDERTAIL COVERTS

VENT

RETRICES (TAIL FEATHERS)

PERIOPHTHALMIC RING (EYE OR ORBITAL RING)

EYE

CHEEK

KEEL (BREASTBONE)

UNDERWING COVERTS

BREAST

BELLY (ABDOMEN)

FEET

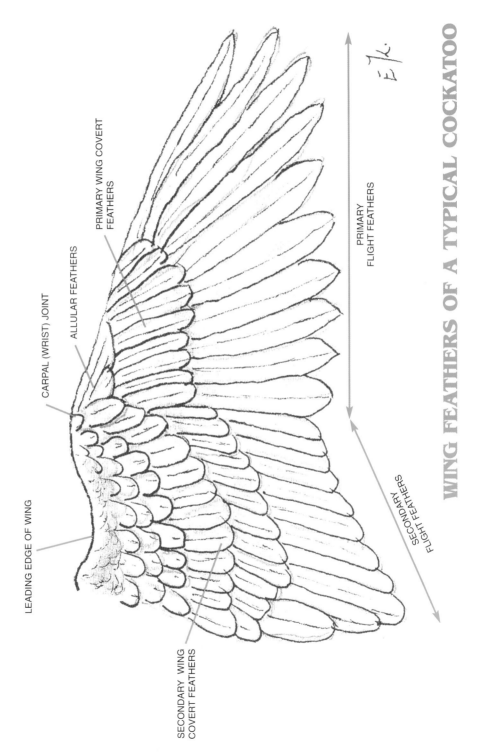

PRIMARY WING COVERT FEATHERS

ALLULAR FEATHERS

CARPAL (WRIST) JOINT

LEADING EDGE OF WING

SECONDARY WING COVERT FEATHERS

PRIMARY FLIGHT FEATHERS

SECONDARY FLIGHT FEATHERS

WING FEATHERS OF A TYPICAL COCKATOO

Simply the best publications on pet & aviary birds available ...

AUSTRALIAN BIRDKEEPER MAGAZINE

Six glossy, colourful and informative issues per year. Featuring articles written by top breeders and avian veterinarians from all over the world.

SUBSCRIPTIONS AVAILABLE

For subscription rates and FREE catalogue contact **ABK Publications**
- details on page 112

Handbook of *Birds, Cages & Aviaries*

This handbook provides a complete overview to the selection, keeping, management and care of both pet and aviary birds from individual pets to larger aviary complexes. Topics include Choosing your Bird, Choosing and Keeping Pet Birds, Housing and Keeping Aviary Birds, Aviary Design, Construction and Management, Plantscaping your Aviary, Nutrition and Feeding, Breeding and Husbandry, General Management and Health and Disease Aspects. A must for the novice and serious aviculturist and all pet bird owners.

■ How to Care For & Tame Your Budgerigar

This easy to read title ia a must for anyone, any age, contemplating purchasing a budgie as a pet.
48 pages - full colour throughout
Covers
- Choosing your pet
- Feeding
- Housing
- Finger Taming
- Taking your bird out of its cage
- Teaching your budgie to talk
- And Much Much More...

The Acclaimed 'A Guide to ...' range

Concise, informative and colourful reading for all bird keepers and aviculturists.

- A Guide to Gouldian Finches
- A Guide to Australian Long and Broad-tailed Parrots and New Zealand Kakarikis
- A Guide to Rosellas and Their Mutations
- A Guide to Eclectus Parrots
- A Guide to Cockatiels and Their Mutations
- A Guide to Pigeons, Doves and Quail
- A Guide to Lories and Lorikeets
- A Guide to Basic Health and Disease in Birds
- A Guide to Australian Grassfinches
- A Guide to Neophema and Psephotus Grass Parrots and Their Mutations (Revised Edition)
- A Guide to Asiatic Parrots and Their Mutations (Revised Edition)
- A Guide to Incubation and Handraising Parrots
- A Guide to Pheasants and Waterfowl
- A Guide to Pet and Companion Birds

ABK Publications stock a complete and ever increasing range of books and videos on all cage and aviary birds. For Free Catalogue and Price List see details on page 112.

PUBLICATIONS

Publishers Note

This title is published by **ABK Publications** who produce a wide and varied range of avicultural literature including the world acclaimed **Australian Birdkeeper** magazine - a full colour, bi-monthly magazine specifically designed for birdlovers and aviculturists. It is the intention of the publishers to produce high quality, informative literature for birdlovers, fanciers and aviculturists alike throughout the world. It is also the publishers' belief that the dissemination of qualified information on the care, keeping and breeding of birds is imperative for the total well-being of captive birds and the increased knowledge of aviculturists.

Nigel Steele-Boyce

Nigel Steele-Boyce
Publisher/Editor-In-Chief
ABK Publications

For further information or Free Catalogue contact:

ABK Publications
P.O. Box 6288
South Tweed Heads
NSW 2486 Australia

Phone: (07) 5590 7777 Fax: (07) 5590 7130
Email: birdkeeper@birdkeeper.com.au
http://www.birdkeeper.com.au